VIRUSES AND TRACE CONTAMINANTS IN WATER AND WASTEWATER

Seminar on Viruses and Trace Contaminants in Water and Wastewater, University of Michigan, 1977

VIRUSES AND TRACE CONTAMINANTS IN WATER AND WASTEWATER

Edited by

Jack A. Borchardt
Chairman, Department of Sanitary Engineering
College of Engineering
University of Michigan
Ann Arbor, Michigan

James K. Cleland
Sanitary Engineer
Michigan Department of Public Health
Lansing, Michigan

William J. Redman
Chemist
Michigan Department of Public Health
Lansing, Michigan

Gordon Olivier
Sanitary Engineer
Michigan Department of Public Health
Lansing, Michigan

230 Collingwood, P. O. Box 1425, Ann Arbor, Michigan 48106

Library of Congress Catalog Card No. 77-076910
ISBN 0-250-40170-3

Manufactured in the United States of America

PREFACE

This volume is a compilation of papers which were presented at the Seminar on Viruses and Trace Contaminants in Water and Wastewater presented by the University of Michigan College of Engineering, Division of Sanitary and Water Resources Engineering, January 26, 27 and 28, 1977. This seminar was a cooperative effort with the Michigan Section of the American Waterworks Association, the Michigan Water Pollution Control Association and the Michigan Department of Public Health.

These agencies have had a long history of cooperation in the continuing education of their respective memberships. This type of seminar, held about every two years, begins as a grass roots request of operators and engineers in the water and wastewater industry for further knowledge in a given area. Requests are received by the respective Education Committees which then designate a seminar planning committee. This committee in cooperation with the University of Michigan plans the seminar, selects the speakers and their topics and finally arranges a two- or three-day program covering the needs of the profession as they may interpret them.

The planning committee did an excellent job in arranging the program, and the speakers responded in a superlative and unselfish fashion in handling their subject matters. The overall message of the conference came through loud and clear. In essence, the understanding of these subjects is far from complete. Progress is being made, but there is still an exceedingly long way to go. It appears that each step forward takes the combined efforts of many technological experts, each interacting with the others. It is this interaction which is one of the vital ingredients in continuing progress. This type of seminar enhances and augments this interaction. The sincere appreciation of all conference organizers is expressed to all the speakers for their important contributions to the success of this effort.

Jack A. Borchardt

INTRODUCTION

John E. Vogt, P.E.

Chief, Bureau of Environmental and
Occupational Health
Michigan Department of Public Health
Lansing, Michigan

It is timely and fitting to acknowledge the great success of past high-level seminars offered at the University of Michigan approximately every three years. It is particularly timely to pay tribute to Jack Borchardt who has been the guiding force behind each course and has offered pertinent subject matter such as coagulation, filtration, taste and odor control, and now, trace contaminants and viruses. These efforts are invaluable to the water and wastewater industries in Michigan and to their leading organizations–the Michigan Section of the American Water Works Association (AWWA) and the Michigan Water Pollution Control Association. The water and wastewater industries salute the efforts of Jack Borchardt in providing a forum where the basic issues and concerns of the operators can be discussed and resolved.

In recent years, we have experienced a seemingly never-ending series of national alarms in the wake of discovery of specific pollutants or contaminants in our environment. Let us review the leaders:

1. Pesticides such as DDT were discovered in trace quantities throughout the environment, and negative effects on the reproductive capacities of certain birds and animals led to national regulatory action.
2. Asbestos-like fibers were found in the drinking water supply of Duluth, Minnesota, and were implicated by some in the development of certain human cancers. I stress "implicated" as there has been little, if any, proof of a real public health concern with this naturally occurring material. Dr. Williams will elaborate on this subject in Chapter 13.
3. Mercury was identified in the food chain in the St. Clair River, resulting in fishing bans and establishing standards of mercury levels in fishing and drinking water. Subsequently, many other elements in the group

known as heavy metals have inspired study and speculation on environmental effects, both to man and especially to aquatic life.

4. Polychlorinated biphenyls (PCB) were discovered in the environment resulting in regulatory action and ongoing studies of environmental effects. PCBs are one example of the thousands of man-made compounds introduced to the world each year, then introduced to the environment through waste discharge or accidental loss.
5. Polybrominated biphenyls (PBB) were introduced into the food chain in Michigan resulting in the slaughter of thousands of cattle, hogs and chickens and requiring a long-term epidemiological study of people, which is just now well underway.
6. The most unlikely candidate for an environmental alarm due to the national concern in 1975—chlorine. Chlorine in combination with certain complex organic substances, many of these naturally occurring, produces compounds which have been shown to be potentially dangerous to various life forms.

Let me urge here a measure of caution—today, because of the advancement of laboratory instrumentation, we are able to find and measure minute concentrations of substances; therefore, let us not eliminate or rule out the use of each compound or element we find in the environment based on speculated damage. How soon we forget the overwhelming public health benefits of DDT in the control of malaria, the industrial benefits of mercury and other heavy metals, and most importantly, the public health benefits of chlorinating drinking water and disinfecting municipal wastewater.

It is absolutely essential to proceed with caution and discretion, and without emotion. We must evaluate relationships such as cost vs effectiveness and risk vs benefit before making regulatory decisions. We cannot abandon a procedure or the use of a chemical that has met the test of time, until thorough research has demonstrated unacceptable risk to the public, or cost-effective alternatives are proven superior. In many instances there is no demonstrated risk at all—simply questions raised by overenthusiastic researchers and ardent environmentalists.

The intent of this book is to acquaint water and wastewater operators, consulting engineers, and the interested public with the basic information concerning trace organics, trace inorganics and viruses. We are not as concerned here with the details of the many research efforts in progress, as much as with the real significance of these substances that have commanded national attention.

We want the reader to be able to address the questions posed by the public on the safety of their drinking water and the adequacy of their wastewater treatment while not ignoring the potential benefits of alternate techniques, or the potential danger of certain substances in the environment. It is simply a matter of proper perspective, balance and objectivity.

Water quality standards and wastewater effluent limitations are being proposed or established by state and federal regulatory agencies. What is the basis for these standards? What are the environmental trade-offs? Hopefully, this book will provide insight into these vital issues. The contributors to this book include regulatory people on the state and federal level who can address this subject.

Many seminars across the nation have been completed or are planned for the future to deal with these subjects. This book, which emanates from a seminar, is designed for the man in the trenches or on the prow of this national effort to protect the health of people and assure a quality environment, who is faced with answering the questions of a concerned and alarmed public.

John E. Vogt, P. E.

Proceedings of the *Viruses and Trace Contaminants in Water and Wastewater* Conference held January 1977, The University of Michigan, Ann Arbor, sponsored by:

The University of Michigan, College of Engineering

The Michigan Section of the American Water Works Association

The Michigan Water Pollution Control Association

The Michigan Department of Public Health

PLANNING COMMITTEE

Wayne H. Abbott, Jr.
Utilities Department
City of Ann Arbor, Michigan

Stuart H. Bogue
Pate, Hirn & Bogue
Consulting Engineers
Southfield, Michigan

Jack A. Borchardt
Chairman, Department of
Sanitary Engineering
College of Engineering
University of Michigan
Ann Arbor, Michigan

Robert E. Hansen
Superintendent of Water
City of Mt. Clemens, Michigan

Gordon E. Jones
Wilkins & Wheaton
Consulting Engineers
Kalamazoo, Michigan

William J. Redman
Chemist
Michigan Department of
Public Health
Lansing, Michigan

James K. Cleland
Sanitary Engineer
Michigan Department of
Public Health
Lansing, Michigan

CONTENTS

SECTION I

VIRUSES

1

THE NATURE OF VIRUSES

Dale R. Stringfellow, Ph.D.
Research Associate
Experimental Biology
The Upjohn Company
Kalamazoo, Michigan

INTRODUCTION

Many important infectious diseases are caused by viruses. Some, such as rabies, small pox, hepatitis and various forms of encephalitis are important because they are frequently fatal; others because they are extremely contagious and create widespread discomfort and loss of work time. Among these are influenza, the common cold, mumps, chicken pox and other respiratory and gastrointestinal disorders. In recent years, viruses have also been implicated in the etiology of certain debilitating central nervous system maladies (multiple sclerosis) and various forms of cancer. It is appropriate, therefore, that a great deal of interest be placed upon determining the epidemiology, pathogenesis and means of controlling viral diseases.

PROPERTIES

Viruses are a heterogeneous class of agents. They vary in size and morphology, and in chemical composition, host range and effects on their host. However, there are properties that all viruses hold in common; these are summarized in Table I. Viruses are extremely small, obligate, intracellular parasites. They contain no cells and, therefore, are completely dependent upon a living cell for their replication and survival. Unlike other living organisms, viruses contain only one type of genetic information.

That is, they contain either RNA or DNA but never both. The viral genome (nucleic acid) is enclosed in a protective protein coat, which is frequently enclosed in an envelope containing both protein and lipids. Viruses multiply only inside the cell. They are absolutely dependent on the host cell's synthetic and energy-yielding apparatuses for their replication and survival. Viruses, therefore, could be visualized essentially as elements of nucleic acid enclosed in a protein coat, which enter the cell and code for production of proteins essential to replication of the nucleic acid and structure proteins necessary for construction of intact virions.

Table I. Properties of a Virus

1. Small obligate intracellular parasite.
2. Contains no functional ribosomes, mitochondria or other cellular organelles.
3. Genetic information is made up of either RNA or DNA, not both.
4. Viral genome is enclosed in a protective protein coat, which may be associated with carbohydrates or lipids of cellular origin (envelope).

Based upon these unusual properties, an age-old debate has centered around whether viruses are actually living organisms. When the virus exists outside the mammalian cell, it obviously has no metabolic processes or apparatuses for replicating itself. However, inside the cell a virus utilizes the cell's own metabolic machinery toward its own end. Perhaps, as Joklik (1972) has suggested, viruses should be called either functionally active or inactive rather than living or dead.

STRUCTURE

The basic structure of a simple enveloped or naked virion is schematically illustrated in Figure 1. Animal viral nucleic acids are composed of either DNA or RNA. Some are double-stranded, others single-stranded, and can exist in either circular conformations. The nucleic acid is composed of repeating subunits arranged in precisely defined patterns. These subunits are called capsomers. Capsids not only protect the viral genome from potentially destructive agents in the extracellular environment, but also play an important role in the ability of the virus to specifically attach to the cell surface, which is a prerequisite for viral entry into the cell. Capsids of animal viruses exist in three basic configurations; helical, cubic or compound symmetry. In helical symmetry, the nucleic acid helix of of the virus is surrounded by protein molecules arranged to form a tubucle structure with a single rotational axis. Such a structure is illustrated in Figure 1 as an enveloped virion. In the second pattern of virus structure, the nucleic acid is condensed and forms the central portion while the capsomers are arranged peripherally with icosahedral symmetry. That is, they are arranged to form 20 equilateral triangular faces constructed in a perfectly symmetrical cubic configuration.

In most viruses, the nucleocapsid is enclosed by an envelope, which is acquired as the virus buds or is released from the cell membrane. Most envelopes consist of lipoproteins in which spikes, made of virus-specific glycoproteins, are attached in a regularly repeating manner and are referred to as peplomers.

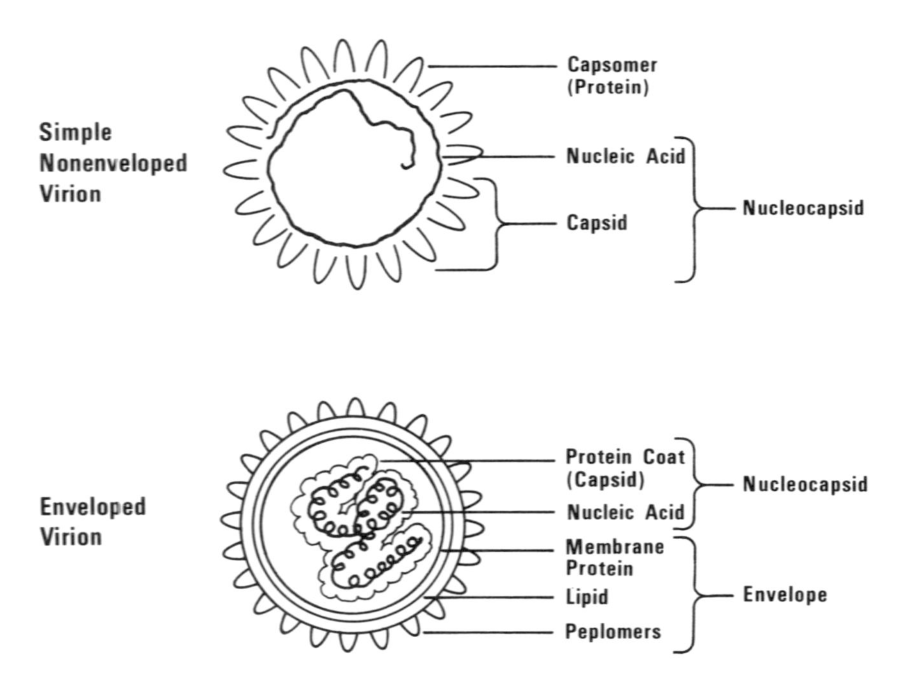

Figure 1. Diagram of the structure of a simple nonenveloped virus with an icosahedral capsid, and an enveloped virus with a tubular (helical) nucleocapsid.

CLASSIFICATION

As summarized in Table II, viruses can be classified according to their morphology, physical and chemical nature. The primary consideration in

Table II. Criteria Used in Classification of Viruses

1. Nucleic acid core (DNA or RNA).
2. Structure (cubic, helical, complex).
3. Enveloped or naked.
4. Site of capsid assembly (cytoplasm or nucleus).
5. Number of capsomers per virion.
6. Size (diameter or diameter x length).
7. Molecular weight of nucleic acid.

dividing all animal viruses into two major groups is whether the virus contains DNA or RNA. Next, the structure of the virus—whether cubic, helical, or complex symmetry—further subdivides the DNA and RNA viruses. Members of each group are subdivided further based on whether they are enveloped or naked. Each of these subgroups can be further divided based upon the site of assembly of the capsid, the number of capsomers per virion, the size of the intact virus particle and the molecular weight of the nucleic acid. In Tables III and IV the classification of animal DNA and RNA viruses is presented. This represents a modification of Melnick's scheme (1971), and is based upon the criteria summarized in Table II. Subgroups of each major virus group have been established based on nucleic acid hybridization and immunologic studies. Some viruses have proven

Table III. Classification of Animal–DNA Viruses[a]

Nucleic acid	DNA				
Structure	Cubic				Complex
Envelope	-			+	?
Site of capsid assembly	Nucleus			Nucleus	Cytoplasm
Number of capsomers	12 or 32	42 or 72	252	162	-
Size (μ)	18-24	40-55	70-80	110	230-300
Nucleic acid molecular weight (10^6)	1.4	2-4	23	40-84	160-240
Group	Parvo	Papova[b]	Adeno	Herpes[c]	Pox[d]

[a]Modification of Melnick (1971).
[b]Includes human wart virus, polyoma and papilloma viruses.
[c]Includes Herpes virus hominis type I (cold sores and fever blisters) and type II (venereal lesions), varacella zoster virus (chicken pox) and cytomegalovirus.
[d]Includes vaccinia virus and variola virus (smallpox).

Table IV. Classification of Animal–RNA Viruses[a]

Nucleic acid	RNA								
Structure	cubic		helical			unknown			
Envelope	-		+			+			
Site of capsid assembly	cytoplasm		cytoplasm			cytoplasm			
Number of capsomers	32	92							
Size (μ)	18-30	54-75	80-120	100-300	60-225	8-120	100	35-40	110-130
Nucleic acid molecular wt (10^6)	2	10	2.5	8	3.4	?	10-12	2	?
Group	Picorna[b]	Reo[c]	Orthomyxo[d]	Paramyxo[e]	Rhadbo[c]	Corona[g]	Oncorna[h]	Toga[i]	Arena[j]

[a]Modification of Melnick (1971).

[b]Entero and rhino viruses (common cold).

[c]Colorado tick fever.

[d]Influenza viruses.

[e]Measles, mumps.

[f]Rabies virus.

[g]Human respiratory viruses.

[h]Animal leukosis and tumor viruses.

[i]Rubella (measles) and Arbo viruses (encephalitis).

[j]Lassa fever virus.

very difficult to work with, and therefore have yet to be classified using this system. For example, not enough is known about the structure or the physical-chemical properties to warrant classification of hepatitis A and B viruses into any of the groups listed.

REPLICATION

Some knowledge of the mechanism of viral multiplication is needed as a basis for understanding how viruses are transmitted, how they cause disease, and how they might be controlled. Viruses and cells are brought in contact in a random fashion, but attachment occurs only if there is an affinity between the surface of the virus and the cell (Figure 2). The plasma membrane of the cell contains specific viral receptor sites which are complementary to attachment sites on the surface of the virion. Apparently, viral attachment is specific for these regions. This is exemplified by the fact that viruses, which can infect specific cell types, for example human cells, in many cases cannot attach to and effectively replicate in murine cells, since they lack specific attachment sites. Once the virion has attached to the animal cell, it must be engulfed by pinocytosis. After engulfment, the virion is enclosed in a phagocytic vacuole and the protein coat removed. At that point the nucleic acid of the virus is released into the cytoplasm and itself functions as messenger (m) RNA or initiates transcription of new mRNA, which specifically codes for early proteins required for the replication of viral nucleic acid. Once early proteins, *i.e.*, polymerases, have been produced, the viral nucleic acid (DNA or RNA) is replicated. Viral nucleic acid (RNA) or new mRNA, which specifically codes for late-appearing structural proteins, is then translated.

The resulting structural proteins are assembled to form capsomers, and the capsid is formed around the nucleic acid of the virion. Intact virus particles are then released from the cell membrane by budding or accumulating until the cell undergoes autolysis, ruptures, and intact virus particles are released. Those virions released by budding are generally enveloped, whereas those released by lysis are in many cases naked. Interestingly, the envelope surrounding virus particles not only contains normal cell membrane components, but also consists of specific viral peplomers previously coded for by the replicating virus. This obviously represents an oversimplification of a very complex phenomenon, the details of which are only beginning to be elucidated.

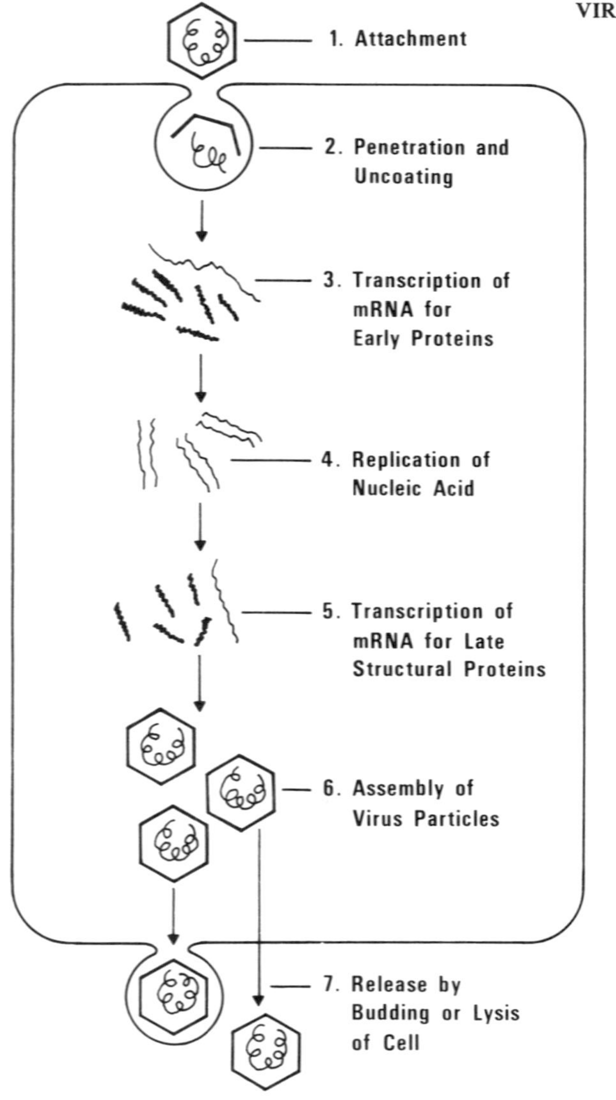

Figure 2. Simplified diagram of the virus multiplication cycle. After attachment at specific receptor sites (1), the intact virion enters the cell by pinocytosis. The protein coat is removed inside a phagocytic vacuole (2) and the viral nucleic acid enters the cytoplasm. Depending on whether the virus contains RNA or DNA, new messenger RNA (mRNA) is produced or the virus' own RNA (3) serves as the template for production of early proteins necessary for replication of viral nucleic acid (4). New mRNA or viral RNA then specifically codes for late structural proteins (5), which are assembled to form intact nucleocapsids (6). Virus particules are then released by budding or accumulate until cell lysis (7).

DISEASES

At the molecular level we have dealt with so far, viruses behave quite differently from other infectious agents. This distinction is not as clear when we consider the effect of viruses on the organism as a whole. Both the pathogenesis and the epidemiology of viruses have much in common with many bacterial and protozoan infections. Before exploring viral pathogenesis at the level of effect on the organism as a whole, some consideration should be given to the effect of viruses at the cellular level. Once a cell is infected with a virus one of three effects may result. First, the cell might be destroyed by the cytocidal or cytopathic action of the virus. Second, a steady-state or persistent virus infection might result, in which case the cell is not destroyed but virus is continually shed or continually present in the cell. The third effect produced by certain oncogenic viruses is that the cell becomes transformed, gaining new morphologic and macromolecular synthesizing properties. All the changes described for virus-infected cells occur in the intact animal. Rather than dealing with a single-cell population, however, destruction or alteration of the cells in the intact animal may affect the function of vital tissues and organs and consequently disrupt normal physiologic functions.

A schematic illustration of various forms of viral infections (pathogenesis) is presented in Figure 3. The virus enters the host at an initial site and generally replicates in cells locally. This local replication may result in a disease state. For example, persons infected with vaccinia virus develop pustules at the site of inoculation. Virus may also be transported by the blood or lymphatic systems or by direct nerve route spread to target organs and tissues in the intact animal. In cells of target organs, the virus again replicates and destroys or alters cellular functions, resulting in disease. An example of this type of infection is poliovirus, which enters the host by way of the alimentary tract, replicates in the gastrointestinal lining, is spread by the blood stream to the central nervous system and replicates in the central nervous system, in some cases causing paralysis and death. Virus produced at the local site, and spread via the blood and lymphatic systems, however, might also infect the secondary organs such as the reticuloendothelial system (spleen and liver). At this site, the virus might again replicate and be spread by way of the blood stream to other target organs such as the central nervous system. Disease might result from the replication of the virus in the reticuloendothelial system, such as in infectious mononucleosis or cytomegalovirus infection. In each case, infection is initiated at a local site and produces recognizable disease in those tissues or is spread by several routes to other as yet uninfected target organs, causing cellular destruction and alteration that results in development of clinical signs and symptoms of illness.

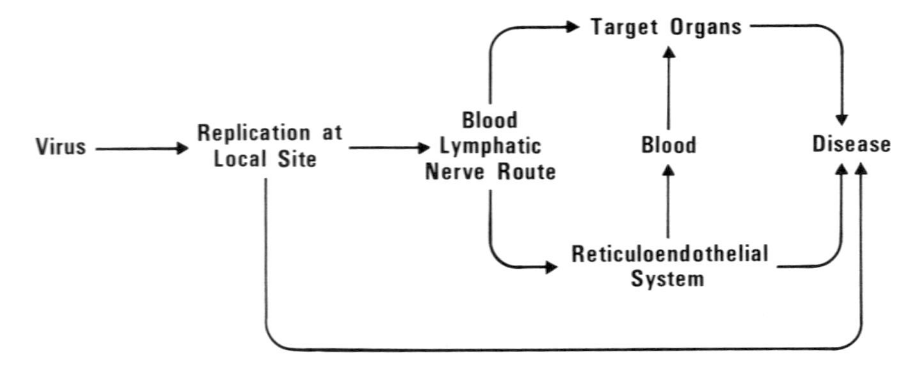

Figure 3. Simplified illustration of the pathogenesis of acute viral infections (modified from Glasgow, 1970).

ALIMENTARY TRACT INFECTIONS

Most viruses never cause primary infection of the gastrointestinal tract because they cannot survive the acid environment of the stomach or the bile-containing duodenum. Certain members of the picornavirus group, however, are stable under these conditions and do cause human gastrointestinal tract infections. The picornaviruses comprise the largest and most important group of human pathogens. They are also the smallest RNA viruses and have been divided into two subgroups–the enteroviruses, (those found primarily in the enteric tract), and the rhinoviruses, (those found primarily in the upper respiratory tract and that cause common cold syndromes in humans). The enteroviruses have been further subdivided into polio, coxsackie and echo viruses (Table V). Each initiates infection in the alimentary tract, causes local symptoms of illness that vary from severe to nonclinical signs and symptoms of disease, and may spread to various target organs, including the central nervous system, causing paralysis and death.

Table V. Viruses that Initiate Infection in the Alimentary Tract

1. Picornaviruses
   a. Poliovirus
   b. Coxsackie virus A
   c. Coxsackie virus B
   d. Echoviruses
2. Hepatitis virus
   a. Hepatitis A (infectious hepatitis)
   b. Hepatitis B (serum hepatitis)

Poliomyelitis was first described in England in 1836 and has been one of the most feared virus infections. Its tragic legacy of paralysis and deformity was a familiar sight a generation ago. Today, by contrast, new cases of poliomyelitis are fairly rare due to the Salk and Sabin vaccines. Replication of poliovirus occurs initially within the nasal pharynx and gastrointestinal mucosa. Virus is then spread by way of the bloodstream or lymphatic system to the target organ, in this case the central nervous system, where signs and symptoms of illness develop. The usual incubation period between contact and onset of paralysis is 10-15 days. Virus is spread from nasal pharyngeal excretions and can be detected in the bloodstream from 3-10 days after infection. Fecal excretion of virus can occur for as long as 30-40 days after infection.

The direct chain of poliovirus infection is difficult to trace because most infections are inapparent. However, it is clear that poliovirus is spread mainly by way of the fecal-oral route, and there is no doubt that epidemics result from contamination of water supplies by sewage. In fact, the rate of transmission of poliovirus in urban communities can be related closely to the method of sanitation practiced in the area. In tropical areas where sanitation is poor, infection is acquired during infancy or early childhood when paralytic disease is uncommon. In more developed areas, major portions of the population may reach adulthood without contact with the virus and subsequently are susceptible to the infection under epidemic conditions. Since the introduction of the Sabin attenuated poliovirus vaccine in 1960, the number of susceptible individuals in most highly developed societies has been reduced to a relatively low level. The Sabin vaccine consists of active attenuated strains of poliovirus. It can be administered by mouth, multiplies in the gastrointestinal tract, is shed in the feces of vaccinated people for extended peiods of time and induces a strong local and systemic immune response.

Coxsackieviruses are the causative agent of numerous different clinical entities that range from the typical common cold syndrome to gastroenteritis to severe aseptic meningitis and paralysis, which may be fatal. Coxsackieviruses are worldwide in distribution and man appears to be the only natural host. Spread of infection probably involves direct contact and fecal dissemination. Coxsackieviruses multiply primarily in the pharynx and in the small intestine, and laboratory diagnosis is made on the basis of virus recovery from the feces, throat, nasal pharyngeal washings or occasionally from cerebrospinal fluid.

Like coxsackieviruses, echoviruses cause a number of specific syndromes in both epidemic and endemic proportions. They cause gastroenteritis, aseptic meningitis and respiratory diseases similar to the common cold. As with other enteroviruses, echoviruses multiply in the nasal pharyngeal area or gastrointestinal tract and can be spread by direct contact, droplets or fecal dissemination.

The other members of the picornavirus group, rhinoviruses, are primarily pathogens of the upper respiratory tract and are one of the chief causes of the common cold syndrome. Rhinoviruses apparently do not infect the gastrointestinal tract and are almost never recovered from stool or rectal swabs.

Another group of viruses that can be spread by way of the gastrointestinal tract are hepatitis A and B viruses. Classically, two forms of hepatitis have been described. One may be orally or parenterally transmitted and is characterized by a short incubation period; the other appears after a much longer incubation period and was thought in the past to require parenteral transmission. The short-incubation hepatitis has been termed hepatitis virus A or infectious hepatitis. The long-incubation disease has at various times been referred to as hepatitis virus B or serum hepatitis. Short-term incubation hepatitis is primarily a disease of children with maximum incidence between the ages of 5 and 14. Infection is widespread and transmitted primarily by the fecal-oral route. Short-incubation hepatitis spreads rapidly from person to person. Thus, there is a high risk of infection among secondary contacts in households and within closed communities. Common source exposure outbreaks have been traced to contaminated water supplies, milk or food. Shellfish taken from sewage-contaminated areas may carry and transmit infection.

The long-incubation hepatitis was long thought to be transmissible only by parenteral route. It has now been demonstrated that the agent of long-incubation hepatitis may be acquired by the oral route and that this virus probably exists in infectious forms in feces as well as blood. Both hepatitis A and hepatitis B viruses have proven to be difficult to isolate and characterize and, therefore, are not classified in any of the major virus groups. The difficulty in working with this virus has also impeded an understanding of the exact pathogenesis and mode of transmission. The short-incubation disease, caused by hepatitis A virus, has an incubation period of 15-50 days and is probably excreted in feces during the incubation period as well as the acute phase of infection, which may last for 70 or more days after exposure. The long-incubation or serum hepatitis caused by hepatitis B virus has an incubation of 45-160 days, and it appears that individuals can have protracted or indefinite carrier states and serve as a common source of infection to other susceptible individuals.

VIRUS STABILITY

In general, infection of the intestinal tract is initiated by viruses from feces that are resistant to the acids, bile salts and enzymes that occur in the gut. Although viruses transmitted by the alimentary route may be inactivated in feces dried in the sunlight, they are, in general, more resistant to inactivation by environmental conditions than the respiratory viruses that must spread directly from infected to susceptible individuals. Viruses transmitted by the alimentary route can persist for some period outside the body. Survival of the viruses outside the body can be affected by a variety of environmental factors (Table VI). Viruses differ enormously in heat stability. In general, the enteroviruses are reasonably heat stable with a decrease in infectivity of no more than 2-4 fold during six hours at 37°C. Heat stability is strongly influenced by environmental conditions. Proteins stabilize all viruses to a greater or lesser extent as do metal ions, particularly divalent cations such as magnesium and calcium. Likewise, enteroviruses are very resistant to acid conditions while other viruses, such as rhinoviruses, are very susceptible. Most viruses, however, are disrupted under alkaline conditions. All viruses are inactivated by electromagnetic radiation, in particular X-rays or gamma rays and by ultraviolet light. X-Rays inactivate viruses primarily by causing breaks in nucleic acid strands. Ultraviolet radiation also damages nucleic acid; in particular, it causes formation of covalent bonds between adjacent pyrimidine molecules. Since infectivity requires expression of the viral genome, both types of radiation are lethal.

Table VI. Factors Affecting Virus Survival

1. pH
2. Temperature
3. Drying (air)
4. Radiation (sunlight)
5. Chemical Reagents
    a. Detergents (anionic vs cationic)
    b. Protein solvents (*i.e.*, phenol)
    c. Formaldehyde
    d. Lipid solvents
    e. Enzymes (*i.e.*, phospholipases)
6. Stabilizing organic matter
7. Cation stabilization

Since the viral genome is protected from the outside environment by its protein coat, agents that disrupt the structure of the capsid can affect viral survival. Anionic detergents, for example, not only solubilize viral envelopes but also dissociate capsids into their constituent polypeptide chains. Protein solvents such as guanidine, urea and phenol also extensively dissociate the viral capsids into their constituent molecules. Formaldehyde is unusual in that it destroys infectivity without significantly affecting antigenicity and, therefore, has been used extensively to prepare inactivated virus vaccines. It destroys infectivity by reacting with amino groups of adenine, guanine and cytosine, which are not involved in hydrogen bond formation. Viruses containing single-strand nucleic acid are, therefore, inactivated readily while those which contain double-strand nucleic acid are more resistant. Lipid solvents such as ether and alcohol and some enzymes such as phospholipases can inactivate enveloped virions by hydrolyzing the lipid-containing envelope. Also, ribonucleases can destroy viral nucleic acid once the protein coat has been removed.

The rate of inactivation of viruses is greatly affected by the presence of stabilizing organic matter and the presence of various types of metallic ions, particularly divalent cation such as magnesium and calcium. Therefore, measurement of inactivation time and stability of viruses should be done under carefully standardized conditions.

INDICATORS OF FECAL CONTAMINATION

As more emphasis has been placed upon the hazards of viruses as contaminants of water resources, attention has been turned toward identifying viruses that can be used to follow human fecal contamination. Table VII presents the four major properties that a virus must meet to be used as an indicator organism. The virus must be relatively stable under natural conditions. That is, the virus should not be unrealistically stable or labile under naturally occurring conditions. The virus should have a high frequency in human populations and be shed in fecal matter. For practical use, the virus should be avirulent and not represent a hazard to employees. Obviously, the virus must be easily detected using routine methods. Viruses are listed in Table VIII. Obviously, the enteroviruses would be ideal candidates since they are present in high frequencies in human populations and represent the major group of viruses spread by the alimentary tract. Coxsackie, echo and wild-type polioviruses are pathogenic and might not be applicable in this type study due to hazards to employees. It should be emphasized, however, that each is present in sewage, and attempts to isolate an avirulent indicator virus would, in some cases, also result in isolation of such pathogens and, therefore, proper training and precautions by employees should be taken.

Table VII. Properties a Virus Should Have to be Used as an Indicator of Human Fecal Contamination

1. Relatively stable under "natural" conditions.
2. High frequency in human population.
3. Low virulence for employee safety.
4. Sensitive detection methods must be available.

Table VIII. Viruses that Might be Considered for Use as Indicator Organisms

Virus	Advantages	Disadvantages
1. Poliovirus (attenuated vaccine strain)	a. Relative stable–nonenveloped virion b. Shed in feces of vaccinated persons. c. Can be detected by routine methods d. Attenuated strain–hazards low	
2. Hepatitis virus	a. Stability? b. Shed in feces of infected persons	a. No reliable detection methods b. Infectious–hazards extremely high

The attenuated vaccine strains of poliovirus are potential candidates for use as indicator organisms. They are relatively stable nonenveloped virions, are shed from the feces of all vaccinated persons for extended periods of time and can be detected by routine methods. Hepatitis virus, on the other hand, is not an organism that could be used to follow human fecal contamination. Hepatitis virus is very difficult to work with. Its stability under natural conditions is unknown and there are no reliable detection methods that could be used in a routine manner. Furthermore, it is infectious and therefore represents a major hazard to those working with it. Based upon these considerations, the attenuated vaccine strains of poliovirus might make the most feasible virus for use in a routine monitoring program. It should be reemphasized, however, that proper education and training of employees should be a prerequisite for such studies.

ISOLATION AND IDENTIFICATION METHODS

Viruses are obligate intracellular parasites and cannot replicate in any cell-free medium. Further, some viruses are fastidious about the kind of cells they infect and replicate in. Most viruses, however, can be grown in cultured cells, embryonated eggs or in laboratory animals, and the cultivation of viruses in experimental animals or cultured cells is an essential prerequisite for their isolation and subsequent study. Methods used in isolation and identification of viruses are summarized in Table IX. More than a half century ago, human and animal cells and organs were cultured *in vitro.* Since Enders *et al.* (1949) reported that poliovirus could be grown in cells cultured *in vitro,* hundreds of previously unknown viruses have been isolated and identified. The growth of many viruses in cell culture can be monitored by the ability of the virus to lyse (destroy) or alter the cell in some fashion, whereby viral infectivity can be measured. Many viruses kill the cell in which they multiply so that infected cell monolayers gradually develop histological evidence of cell damage or cytopathic effect. By use of agar-containing media, the cytopathic effect of virus infection can be localized in specific areas of a cell monolayer. Cells only in that particular area are destroyed, creating areas of dead cells. These plaques can be counted to enumerate the number of virus particles present.

Table IX. Animal Virus Isolation and Identification Methods

1. Mammalian tissue cultures (organ or cell)
   a. Cytopathic effect (CPE)
   b. Plaque formation
   c. Hemadsorption
   d. Interference
2. Embryonated eggs
3. Laboratory animals
4. Hemagglutination
5. Electron microscopy
6. Immunologic evaluation
   a. Virus neutralization
   b. Hemagglutination inhibition
   c. Gel diffusion
   d. Complement fixation
   e. Immunofluorescence
   f. Radioimmunoassay

In many cases, viruses that bud from the cytoplasmic membrane of the cell acquire the ability to specifically adsorb or attach to erythrocytes. The phenomenon known as hemadsorption is due to the incorporation into the plasma membrane of newly synthesized viral proteins that have an affinity for red blood cells. Hemadsorption can be used to recognize infection with noncytopathic viruses. The multiplication of one virus in a cell often inhibits the multiplication of another virus that subsequently may enter the cell. This phenomenon, known as interference, can also be used to identify viruses.

Prior to the 1950s when cell cultures really began to make an impact on virology, the standard host for cultivation of many viruses was the embryonated hen's egg. Nearly all viruses known at that time could be grown in cells of one or another of the embryonic membranes. Embryonated eggs are rarely employed for virus isolation now, but due to production of such high yield of certain viruses such as influenza virus, this system is used both for research studies and vaccine production. In the past, laboratory animals have also been used in the isolation and identification of a number of viruses. Like embryonated eggs, however, laboratory animals are infrequently used in diagnostic laboratories since cell cultures are so much easier to handle and are much more flexible.

A number of laboratory methods are presently available for the enumeration of viruses, including the plaque assay previously described and the hemagglutination assay, which is based upon the ability of viruses to adsorb red blood cells of various animal species. Each virus particle is multivalent; that is, it can adsorb to more than one red blood cell at a time, and a lattice is formed which settles out in a highly characteristic manner readily distinguishable from the pattern exhibited by unagglutinated cells. Electron microscopy has also been used in the identification and isolation of virus particles.

Serologic surveys of populations of vertebrates are widely used in epidemiological studies and provide information about the distribution and spread of viruses in time and space. In diagnostic virology, antibodies of known specificity are used to identify viruses recovered from diseased or healthy individuals. In laboratory research, serological methods provide the potential for recognizing different antigenic components of a virus and for measuring the concentration of such antigens. Many different available serological techniques are summarized in Table IX. Certain antibodies interact with virions and neutralize their infectivity. Antibodies can inhibit virus-mediated hemagglutination by blocking the particular antigens on the surface of the virion and are responsible for the phenomenon. Antigen-antibody interaction can be detected by observing precipitation reactions in semisolid gels, and antigen-antibody complexes will fix complement and can be used in the standard complement fixation procedure. Immunofluorescence is a technique used not only to detect infectious virus particles but can be used to detect cells that have been infected with viruses by the specific interaction of an immunoglobulin which has been tagged with a dye such as fluorescein (fluorescent antibody) with a specific viral antigen. The availability of carrier-free preparations of radioisotopes (especially iodine), which can be readily coupled to tyrosine residues of proteins has led to the development of radioimmunoassays for analyzing either antigens or antibodies. These procedures are very sensitive and accurate and are relatively simple to perform.

SUMMARY

This chapter briefly covers the nature of viruses and virus infections, their importance as human pathogens and the potential hazards they might impose in our environment. The enteroviruses and hepatitis viruses are the principal etiological agents spread through the alimentary tract and represent potential hazards to our water resources. As further understanding is achieved on the nature of other diseases and as other viruses are isolated and identified, this list may be expanded. This area is obviously in its infancy but may prove critical to the preservation of our environment and the survival of man.

REFERENCES

Enders, J. F., T. H. Weller and F. C. Robbins. *Science* 109:85-77 (1949).
Glasgow, L. A. *J. Gen. Physio.* 56:212-226 (1970).
Joklik, W. K. in *Microbiology*, W. K. Joklik and D. T. Smith, Eds. (New York: Appleton-Century-Crofts, 1972), pp. 713-724.
Melnick, J. L. *Prog. Med. Virol.* 13:462-484 (1971).

For Further Reading

Melnick, J. L. "Enteroviruses," in *Viral Infections of Humans,* A. S. Evans, Ed. (New York: Plenum Medical Book Co, 1976).
McCollum, R. W. "Viral Hepatitis," in *Viral Infections of Humans,* A. S. Evans, Ed. (New York: Plenum Medical Book Co., 1976).
Evans, A. S. "Epidemiological Concepts and Methods," in *Viral Infections of Humans,* A. S. Evans, Ed. (New York: Plenum Medical Book Co., 1976).
Fenner, F., B. R. McAuslan, C. A. Mins, J. Sambrook and D. O. White. *The Biology of Animal Viruses* (New York: Academic Press, 1974).

GLOSSARY

Antigenicity–capability of causing production of antibody.

Capsids–protein covering referring to vision.

Capsomers–structural units.

Cytomegalovirus–Herpes family, produces disease symptoms only in infants (usually in mouth and lips)

Cytopathic–pertaining to a diseased condition of a cell.

Cytoplasmic–substance of a cell exclusive of the nucleus.

Cytotoxic–cytolytic; destructive to cells.

Echoviruses–belong to a large group of viruses causing meningitis, enteritis (with or without fever).

Elute–remove with a solvent.

Eluate–resulting solution.

Enteroviruses–viruses that multiply in the digestive tract, including Coxsackie, polio, etc.

Erythrocytes–mature red blood cells.

Etiological–the study of the causes of disease.

Genome–one haploid set of chromosomes with the genes.

Hemadsorption–substance adsorbed *on* red blood cells.

Hemagglutination–clumping of red blood cells.

Hematopoietic–blood forming organs.

Murine–mouse.

Mononucleosis–viral disease characterized by sore throat, swollen lymph glands; no treatment except cortisone.

Oncogenic–relating to tumor formation.

Phagocytic–a cell capable of ingesting bacteria.

Picornavirus–very small viruses which are insoluble in a lipid solvent such as ether.

Pinocytosis–term for the adsorption of liquids by phagocytic cells.

Pyrimidine–($C_4H_4N_2$) crystalline heterocyclic base.

Reticuloendothelial–dealing with spleen and liver.

Rhinoviruses–nose related.

Serological–study of serum.

Vacuole–a minute space in any tissue.

Virion–a complete virus particle surrounded by a protective coat which serves as the vehicle for its transmission from one cell to another.

2

VIRUS DETECTION METHODS–COMPARISON AND EVALUATION

Peter T. B. Shaffer, Ph.D.

Manager, Development
R & D/Water Management
The Carborundum Company
Niagara Falls, New York

INTRODUCTION

This chapter might better be entitled, "Viruses in Water-Monitoring–Problems and Pitfalls," as it discusses some of the methods for recovering viruses from various waters and some advantages and disadvantages of each. The procedure established in both the field and laboratory by researchers at The Carborundum Company for recovering viruses from large water samples is described and, finally, this discussion presents some of the inherent limitations of virus-seeding experiments.

A number of excellent review articles, such as those by Hill *et al.* (1971) and Sobsey (1976), have been published describing various virus recovery techniques. Similarly, the proceedings of a number of virus-in-water conferences such as those in Mexico City in 1974 (Berg *et al.*, 1976) and Austin, Texas, in 1974 (Malina, 1974), provide in-depth discussions of a number of the specifics of virus recovery from waters.

Before discussing techniques, their capabilities and limitations, we must put the matter into perspective and consider just what the problem of virus detection involves. A typical virus, polio 1 for example, is a roughly spherical particle having a diameter of approximately 25 nanometers (2.5×10^{-6} cm), with a density of 1.34 g/cm^3. This corresponds to a weight of approximately 10^{-17} g. Thus, one virus in a liter of water represents one part in 10^{20} by weight. Compared to organic analyses in the parts per billion range (10^{-9}), the detection of one virus in a liter of water (10^{-20}) would appear to be an insurmountable task. By the direct approach, it is.

What levels of virus contamination are reasonable to expect? Experiments have shown samples of every type of water apparently to be virus-free. Zero is, however, a statistically unattainable lower limit. The approximate upper limits, Carborundum studies have shown, range from 0.01 particle, or plaque forming unit per liter (PFU/liter) in potable water to 8000 PFU/liter in raw sewage (Table I) (Shaffer *et al.*, 1976).

Table I. Maximum Levels of Viruses Demonstrated for Various Types of Water

	Number (PFU/liter)	Approximate Concentration (wt %)
Raw Sewage	8000	10^{-14}
Secondary Effluent	4000	10^{-15}
Chlorinated Secondary Effluent	10	10^{-17}
Surface Waters	0.25	10^{-19}
Finished Drinking Water	0.01	10^{-20}

Since viruses cannot be recovered and measured directly using classical analytical methods, how can their presence be demonstrated and their concentration be determined? Their concentration and manifestation relies first on their preferential adsorption to a variety of surfaces (or other means of concentration), their ability to pass through the usual submicron filters to separate them from filterable bacteria, and their ability to infect living cells and, in so doing, bring about visible cytopathic effects (CPE) in the host cell.

Simply, the charge on the virus particle is controlled by adjusting the chemistry of the water system, and the virus is then adsorbed to a surface of opposite charge. After removal of excess water from the system, the virus is recovered in a small volume of alkaline buffer, which reverses the virus charge and causes it to be repelled. Viruses in the concentrate are allowed to infect living tissue cultures in which CPE can be observed.

All this must be done while minimizing or eliminating interferences from various other agents—chemical, physical or biological—present at concentrations as much as 10^{19} times as great. Each water system is chemically different and, as such, presents a potentially new series of interferences to study. In spite of all of this, under ideal conditions viruses can be detected at concentrations as low as 5 PFU per 1900 liters with over 50% confidence (Hill *et al.*, 1974); and routinely from drinking water at levels of 1 PFU per 150 liters (Jakubowski *et al.*, 1975).

GENERALIZATIONS

Adsorption to a surface and elution is the only presently employed method for the recovery of viruses from large volumes of water. Many other techniques have been and are being used under special cases. It is not possible to separate these methods without overlap, but in general, let us consider the properties of viruses that have been exploited in various recovery schemes. These may be divided generally into five classifications: density, size, charge, adsorption and chemistry. Consider each of these properties and some of the ways they have been utilized.

Density

Viruses are somewhat more dense than water (1.3 g/cm^3). In spite of their small size, they can be completely sedimented by prolonged exposure to forces of more than 100,000 times gravity (Schwerdt, 1965). In practice, debris and macroscopic material are first removed by low-speed centrifugation. Viruses in the supernatant liquid are then removed by prolonged centrifugation at high speeds. Finally, viruses virtually free of most other organic matter can be obtained by fractional sedimentation by centrifugation in liquids of closely controlled densities.

Size

Any solid particle can be removed if a fine enough filter is used. A barrier, such as a dialysis membrane having permeability measured in the Angstrom range (typically 50 Å), will permit water molecules and other low-molecular-weight species to pass, while retaining those whose size is greater than that of the porosity. Unlike ordinary filtration, large driving forces such as high pressure (ultrafiltration, reverse osmosis), are required to drive the liquid through the membrane at a reasonable rate.

Crystallization of water provides another means of virus concentration. Larger particles or molecules of different chemistry are preferentially repelled by the growing ice crystals. Virus-in-water suspensions can be partially crystallized, concentrating the viruses in the last fractions of liquid (Katzenelson, 1976).

Charge

Viruses behave as typical amphoteric protein particles. Under acidic conditions, at pH levels below their isoelectric point, they have a net positive charge, while above the isoelectric point they are negatively charged. Application of electric gradients causes the virus to migrate. This migration has been utilized using suitable permeable barriers to cause preferential fractionation of viruses, and other proteins, into comparatively small volumes.

Similarly, ion exchange resins can be used to remove viruses from large volumes of water. Although virus adsorption and elution is based on a similar charge attraction, this is covered separately.

Adsorption

Viruses are concentrated and recovered from large volumes of water by adsorption to fixed substrates such as microfiber glass papers, or to macroscopic particles, which are then recovered by filtration or sedimentation. Substrates used are glass, or silica in several forms, magnetic iron oxide (Fe_3O_4), clays, aluminum hydroxide floc, ferric hydroxide floc, insoluble polyelectrolytes and a variety of insoluble inorganic compounds.

The chemistry of the water is controlled to develop the proper charge on the virus particles, and polyvalent ions may be added to enhance adsorption; then the virus suspension is passed through any of a number of microporous filters. The recovered virus concentrate, eluted into an alkaline buffer, is placed on cell sheets and cytopathic effects are observed.

The oldest method for virus recovery from large volumes of water uses cotton as the adsorbing medium. Sterile cotton pads are suspended in water streams for extended periods, and the adsorbed viruses are recovered by eluting with such elutants as alkaline beef extract or fetal calf serum.

Removal of viruses by formation of insoluble flocs has been attributed to coprecipitation. Recent studies by Wallis (1976) have shown, however, that virus removal by preformed aluminum hydroxide floc is as effective as that resulting from *in situ* formation of the floc.

Chemistry

In a system composed of two immiscible liquids, each component will establish a concentration gradient between the two liquids based on its partition coefficient—the ratio of its equilibrium concentrations in the two phases. The larger this coefficient, the greater will be the concentration of the component in one phase relative to that in the other. Viruses in water segregate preferentially in aqueous two-phase systems containing dextran and polyethylene glycol in the presence of sodium chloride (Albertsson, 1971, 1974), providing yet another way to concentrate them.

DESCRIPTION OF SEVERAL COMMONLY USED METHODS

Although many laboratories use a wide range of virus-concentrating techniques under special conditions, only three techniques are generally accepted by those in the virology field. These are the adsorption, two-phase and filtration concentration techniques. Of these, only the first seems applicable at present to studies of large-volume water samples (more than 100 liters).

First evidence in the literature that human enteric viruses could be concentrated by adsorption to a surface, dates back more than 45 years to the work of Sabin (1931). In the intervening years, viruses have been concentrated by adsorption to a wide variety of materials, such as aluminum hydroxide, (Schaeffer and Bribner, 1933; Wallis and Melnick, 1967), aluminum phosphate (Miller and Schlesinger, 1955; Wallis and Melnick, 1967), calcium phosphate (Salk, 1941; Wallis and Melnick, 1967), polyelectrolytes (Wallis *et al.*,1969; Wallis and Melnick, 1970), magnetic iron oxide (Warren *et al.*, 1966; Rao *et al.*, 1968), various membranes and filters (Metcalf, 1961; Cliver, 1965; Wallis *et al.*, 1972a, 1972b) and fiberglass (Wallis *et al.*, 1972b).

In practice, if conditions warrant, the water to be sampled may be filtered through nonadsorbing filters such as protein treated cellulose acetate (Wallis *et al.*, 1972b) to remove particulates that might otherwise block the virus-adsorbing filters. The filtered water is then adjusted chemically to optimize virus retention and passed through the virus-adsorbing filters.

Conditions used for virus adsorption depend on the types of viruses being sought and the quality of the water being sampled. In clean water, only a slight acidification is required, but in more highly contaminated waters, the preferred conditions for adsorption of human enteric viruses are pH 4.5 with 0.05 *M* magnesium chloride (Metcalf, 1976), or pH 3.5 with 0.0005 *M* aluminum chloride (Wallis, 1976).

Use of clarifying filters introduces two problems that must not be overlooked. First, the composition of the clarifying filters must be such that viruses themselves will not adsorb. Second, the loss of viruses associated with the filtered particulates must be considered.

After processing the prescribed quantity of water, excess water is removed from the system and the viruses are eluted. Again, depending on the virus, the elution conditions can vary widely. Routinely, an alkaline (pH 9.5-11.5) solution of a buffer (0.05 *M* glycerine or 0.1 *M* sodium bicarbonate), alone or in the presence of proteinaceous matter (3% beef extract or fetal calf serum), is passed through the adsorbing filters. The eluate suspension is immediately neutralized and further reconcentrated.

If clarifying filters are used, it may be desirable to elute them. Surface-associated viruses will be recovered. If the viruses physically entrapped in the particulate material are to be recovered, a procedure such as ultrasonic treatment (Wellings *et al.,* 1976) is required to break up the clumped matter and free the viruses.

Viruses in the neutralized eluates are further reconcentrated by reacidification and readsorption to membrane surfaces, by adsorption to aluminum hydroxide floc, by two-phase separation or by hydroextraction. The first, membrane adsorption, is simply a repeat of the initial concentration on a progressively smaller scale. It should be remembered that each step, while reducing the volume of eluate, becomes increasingly complicated by various interferences.

Certain agents present in many waters will concentrate in parallel with the viruses. These include the acid-insoluble, alkali-soluble humic acids and a number of proteinaceous substances that behave like the viruses. The first yields insoluble precipitates that block membrane filters and entrap viruses. The second, the so-called membrane-coating components (MCC), may reach concentrations that they successfully compete with the viruses for adsorption sites and cause a loss of viruses into the filtrate. The membrane adsorption method is not suitable for reconcentration if the eluate contains beef extract or serum, both of which are MCC.

A second method of reconcentration involves the formation of an aluminum hydroxide floc, to which the viruses are adsorbed and recovered by centrifugation. Although the resulting concentrate is of somewhat larger volume than that obtained by other methods, the method is applicable to a wide range of concentrates. Viruses are recovered from the floc by elution with an alkaline buffer and by recentrifuging. The clear supernatant liquid contains the virus product.

Eluates containing much proteinaceous matter are frequently reconcentrated best by the two-phase separation (Albertsson, 1971, 1974; Shuval *et al.,* 1969). To the virus suspension are added sodium dextran sulfate, polyethylene glycol and sodium chloride. Overnight at 4°C, two immiscible phases form, the lower of which, containing the viruses, is drawn off. Additional sodium chloride is added, and overnight two immiscible phases reform. The viruses are finally recovered from the upper phase.

The last method presently receiving considerable attention is hydroextraction, or dialysis (Cliver, 1967). The sample, usually of the order of 100 ml or less, is placed in a dialysis membrane and covered with a concentrated solution of polyethylene glycol (PEG) or with solid PEG itself. Water, salts and low-molecular-weight materials migrate through the dialysis membrane, while high-molecular-weight compounds and viruses are retained.

After a period of up to one day at 4°C, the highly concentrated virus fraction is recovered from the membrane using small volumes of phosphate buffer, beef extract or serum.

The above procedures have been applied in many combinations; however, the one method found most generally suitable in our extensive field and laboratory experiments in a wide variety of natural waters is the initial use of membrane adsorption followed by floc reconcentration and finally hydroextraction. Samples of up to 4000 liters of water may be reduced to less than 10-ml concentrates. Numerous natural virus isolates from wide varieties of waters have thus been obtained.

ADVANTAGES AND LIMITATIONS

There is no single method of virus concentration applicable to all types of viruses or to all waters. Each case presents its own problems. A method that works well in one set of circumstances may not be applicable to another.

The only generally accepted method applicable to the testing of large volumes (more than 100 liters) in reasonable time is the membrane adsorption technique. Sampling at rates of 4-25 liters/min is practical using apparatus sufficiently portable to be used in the field. An alternate method applicable to processing large volumes of water uses reverse osmosis; however, the apparatus is heavy, not readily portable and water throughput rates are generally less than 1 liter/min. Perhaps the biggest problem with reverse osmosis concentration is that it concentrates, as do all ultrafiltration processes, a wide range of extraneous materials. These at high concentrations present problems of solids precipitation, plugging of membranes and the possibility of virus inactivation or toxicity to the tissue cultures.

To recover viruses from smaller volumes of water, such as from the eluates of the membrane concentrate from a large volume of water, several options exist. Additional membrane reconcentration steps can be used, providing interferences due to floc formation or virus breakthrough due to build-up of MCC are not encountered.

If materials are present that will, at higher concentrations, become viricidal or cytotoxic, one must rule out ultrafiltration and dialysis and resort to separations based on floc adsorption, or two-phase separation. On the other hand, if they are not a problem, ultrafiltration is very effective.

By selecting a sequence of different methods for each stage of the virus concentration-reconcentration sequence, one minimizes the chance that interfering agents will concentrate in parallel with the viruses. For example, humic acid precipitates under acidic conditions and redissolves during virus elution with alkaline buffers. Floc reconcentration at near-neutral conditions will not cause the humic acid to reprecipitate and reconcentrate with the viruses.

DISCUSSION AND CONCLUSIONS

Each virus-in-water sampling program must be considered unique. Many waters behave similarly, and only experience teaches how far these similarities can be extended. To cite an example, all drinking waters cannot be treated the same. Some, especially those whose sources are Canadian and northern United States bog and forest country, are rich in humic acids. These acid-insoluble compounds present serious problems and, in fact, preclude the use of membrane reconcentration procedures.

Throughout this chapter, the use of recovery percentages has been intentionally avoided. Publications by several authors will show recovery efficiencies for a single procedure that may range from less than one to more than 100%. While these numbers could be questioned, mentioning them points out that some workers have little or no success with methods that others find highly effective. Whether this is due to subtle differences in the water used, to laboratory techniques or to some other factors, has not yet been determined.

One entering the field of virus-in-water monitoring must also be warned that data obtained under carefully controlled laboratory conditions will likely differ from that obtained under field conditions.

The use of exogenously added viruses has many advantages in the comparison of various techniques and apparatus. It should be remembered, however, that these so-called seeding experiments introduce a degree of artificiality to the experiments and, with this, potential pitfalls.

Seeding with virus is the only way to ensure the presence of sufficient numbers of virus to carry out statistically valid laboratory experiments. Thus, experiments to compare techniques and procedures, to understand the chemistry of adsorption, elution, reconcentration, and to study inactivation and stability must, of necessity, be seeded with comparatively high levels of virus.

Using seeding experiments under field conditions will determine whether the selected procedures, apparatus and conditions are suitable. Care must be exercised to ensure that too much emphasis is not placed on the numbers that result from such seeding experiments. They are a guide, and the results should not be accepted blindly.

Consider several factors that can render seeding experiment data at least suspect. The majority of naturally occurring viruses, and for discussion consider the human enteric viruses only, derive from fecal pollution. As such they are associated, initially at least, with solid or semisolid particulate matter. They may or may not be recoverable by elution from this particulate matter depending on whether they are on the surface or physically entrapped.

Exogenously added viruses are initially added as suspensions in solids-free media. In the test water these viruses will do several things. First, they will tend to disperse in the test water and, interestingly, the resulting concentration is frequently less than that predicted on the basis of laboratory titration of the stock. This apparently is not a toxic effect since the concentration measured shortly after dispersion will then remain effectively constant over extended periods. The initial drop may typically range from one-half log or less (30%) to as much as several logs (99+%). Yet the resulting concentration may then remain essentially constant.

Secondly, the exogenously added viruses, present initially at concentrations of 10^6 or more per milliliter will frequently undergo declumping. This effect tends to counteract the initial drop in titer. The declumping appears to occur as a consequence of adsorption and elution. More viruses may appear to be present in the first eluate than would be predicted on the basis of the original water control.

It cannot be ruled out that these first two factors may be related; however, 10^6 viruses per milliliter (10^{22} molecules of water) into many gallons of test water would not be expected to aggregate simply on the basis of collision probabilities.

Finally, the dispersed viruses will adsorb to particulate matter in the water. Notably, viruses adsorb to clay and various oxides and hydrated oxides such as those of aluminum and iron.

Therefore, it is obvious that seeded viruses, free or adsorbed to the surface of particulates will not behave in exactly the same way as naturally occurring viruses.

Add to this the difference in behavior of the individual viruses themselves. Laboratory strains of viruses frequently behave differently from natural viruses. In fact, each strain of a single virus such as polio 1 has its own characteristics. Effects of temperature of growth rates (T-markers) (Carp and Koprowski, 1962), and rates of inactivation by extreme pH and disinfectants such as chlorine (Shaffer, 1977) all differ among various strains.

Add to this the fact that one strain of one type of virus is usually seeded to demonstrate the behavior, hopefully, of all enteroviruses, and one can see the potential pitfall.

To cite an example, the Mahoney strain of poliovirus 1 replicates at 37.5°C at the same or slightly greater rate after exposure to 40°C ambient while the LSc strain usually fails completely to replicate. To cite another example, in one series of field studies recoveries were inexplicably low. The cause, it developed, was an inadvertent substitution of the Chat strain of poliovirus 1 instead of the designated LSc strain. Studies later showed that more than 90% of the Chat strain was being inactivated during elution at pH 11-11.5. LSc strain is comparatively stable under these conditions.

A good criterion to show that virus-in-water studies are being performed successfully is the recovery of more than one family of viruses (polio, echo, coxsackie A and coxsackie B) and more than one type of virus within a family. Recoveries approximately in the proportions expected, based on the quality of the waters tested, provide further assurance. With these facts in hand, one can feel confident of his field procedures.

In closing, it should be pointed out that failure to recover natural viruses is not necessarily an indication that a method is ineffective. Many waters, even some wastewaters on occasion, are found to yield no natural virus isolates.

REFERENCES

Albertsson, P-A. *Partition of Cell Particles and Macro-molecules* 2nd ed. (New York: John Wiley & Sons, Inc., 1971).

Albertsson, P-A. "Concentration of Virus by Phase Partition," *Virus Survival in Water and Wastewater Systems,* proc. of conference, Austin, Texas (1974), pp. 16-18.

Berg, G. H., L. Bodily, E. H. Lennette, J. L. Melnick and T. G. Metcalf. *Viruses in Water*, proc. of conference, Mexico City (1970).

Carp, R. I. and H. Koprowski. "A New Test of the Reproductive Capacity Temperature Marker of Poliovirus: The Limited Thermal Exposure Test," *Virology* 16:71-79 (1962).

Cliver, D. O. "Factors in the Membrane Filtration of Enteroviruses," *Appl. Microbiol.* 13:417-25 (1962).

Cliver, D. O. "Detection of Enteric Viruses by Concentration with Polyethylene Glycol," in *Transmission of Viruses by the Water Route,* G. Berg, Ed. (New York: Wiley-Interscience, 1967), pp. 109-141.

Hill, W. F., Jr., E. W. Akin and W. H. Benton, "Detection of Viruses in Water: A Review of Methods and Application," *Water Res.* 5:967-995 (1971).

Hill, W. F., Jr. Personal communication to the writer (1974).

Jakubowski, W., W. J. Hill, Jr. and N. A. Clarke. "Comparative Study of Four Microporous Filters for Concentrating Viruses from Drinking Water," *Appl. Microbiol.* 30(1):58-65 (July 1975).

Katzenelson, E. "Virologic and Engineering Problems in Monitoring Viruses in Water," *Viruses in Water*, proceedings of conference, Mexico City (1976).

Malina, J. F., Jr. and B. P. Sagik. "Virus Survival in Water and Wastewater Systems," proc. of conference, Austin, Texas (1974).

Metcalf, T. B., "Use of Membrane Filters to Facilitate the Recovery of Virus from Aqueous Suspensions," *Appl. Microbiol.* 9:376-379 (1961).

Metcalf, T. G. Personal communication to the writer (1976).

Miller, D. H. and R. W. Schlesinger. "Differentiation and Parification of Influenza Viruses by Adsorption on Aluminum Phosphate," *J. Immunol.* 75:155-160 (1955).

Rao, V. C., R. Sullivan, R. B. Read and N. A. Clarke, "A Simple Method for Concentrating and Detecting Viruses in Water," *J. Am. Water Works Assoc.* 60:1288-1294 (1968).

Sabin, A. B. "Experiments on the Purification and Concentration of the Virus of Poliomyelitis," *J. Expl. Med.* 56:307-317 (1931).
Salk, J. E. "Partial Purification of the Virus of Epidemic Influenza by Adsorption on Calcium Phosphate," *Proc. Soc. Expl. Biol. Med.* 45:709-712 (1941).
Schaeffer, M. and W. B. Bribner. "Purification of Poliomyelitis Virus," *Arch. Pathol.* 15:221-226 (1933).
Schwerdt, C. E. "Chemical and Physical Methods," in *Viral and Rickettsial Infection of Man*, F. C. Horsfall and I. Tamm, Eds. (Philadelphia: J. B. Lippincott, 1965).
Shaffer, P. T. B., R. E. Meierer and C. D. McGee. "Isolation of Natural Viruses from a Variety of Waters," AWWA Water Quality Technology Conference, San Diego, California (December 5-8, 1976).
Shaffer, P. T. B. Unpublished data, 1977.
Shuval, H. I., B. Fattal, S. Cymbalesta and N. Goldblum. "The Phase-Separation Method for the Concentration and Detection of Viruses–Water," *Water Res.* 3:225-240 (1969).
Sobsey, M. D. "Methods for Detecting Enteric Viruses in Water and Wastewater," in *Viruses in Water,* proc. of conference, Mexico City, 1974, G. Berg, H. L. Bodily, E. H. Lennette, J. L. Melnick and T. G. Metcalf, eds. (1976), pp. 89-127.
Wallis, C. and J. L. Melnick. "Concentration of Viruses on Aluminum and Calcium Salts," *Am. J. Epidemiol.* 85:(3)459-468 (1967).
Wallis, C., S. Grinstein, J. L. Melnick and J. E. Fields. "Concentration of Viruses from Sewage and Excreta on Insoluble Polyelectrolytes," *Appl. Microbiol.* 18(6):1007-1014 (1969).
Wallis, C. and J. L. Melnick. "Detection of Viruses in Large Volumes of Natural Waters by Concentration on Insoluble Polyelectrolytes," *Water Res.* 4:787-796 (1970).
Wallis, C., M. Henderson and J. L. Melnick. "Enterovirus Concentration on Cellulose Membranes," *Appl. Microbiol.* 23(3):476-480 (1972).
Wallis, C., A. Homma and J. L. Melnick. "A Portable Virus Concentrator for Testing Water in the Field," *Water Res.* 6:1249-1256 (1972b).
Wallis, C., Personal communication to the writer, 1976.
Warren, J., A. Neal and D. Rennels. "Adsorption of Myxoviruses on Magnetic Iron Oxides," *Proc. Soc. Expl. Biol. Med.* 121:1250 (1966).
Wellings, F. M., A. L. Lewis and C. W. Mountain. "Demonstration of Solids-Associated Virus in Wastewater and Sludge," *Appl. Environ. Microbiol.* 31(3):354-358 (1976).

3

THE EFFECT OF UNIT PROCESSES OF WATER AND WASTEWATER TREATMENT ON VIRUS REMOVAL

Joseph E. Malina, Jr.
P.E., Professor & Chairman
Department of Civil Engineering
The University of Texas
Austin, Texas

INTRODUCTION

The survival and transmission of viruses in aquatic and terrestrial systems has been recognized as a potential public health problem. Viruses are the agents of various diseases of man, such as poliomyelitis, hepatitis, gastroenteritis and others. These virus pathogens are introduced into the wastewater system through the excrement of an infected population and more than 100 distinct serotypes of viruses may be present in municipal wastewater throughout the world. The discharge of untreated or partially treated wastewaters and runoff from urban and agricultural areas into surface waters introduce viruses into potential water supplies.

The national policy to eliminate the discharge of pollutants from navigable waters will require more effective treatment of municipal wastewaters. The goals of PL 92-500 call for all municipal wastewater treatment plants to have installed by July 1, 1977, some type of secondary (biological) treatment. Advanced waste treatment processes may be necessary in those situations where reuse of water is considered.

The indirect reuse of municipal wastewater effluent is common practice. The effluent of one municipality soon becomes the influent to the water supply system of the municipality immediately downstream. Therefore, it is even more important that the effectiveness of municipal wastewater treatment systems be understood in regard to the survival or inactivation of viruses.

The processes used in advanced treatment or in treating water supplies are almost the same. The main difference is the type of water processed. In the one case, the effluent from a biological treatment plant enters the advanced treatment system, while surface water from a lake or stream is the influent of the water treatment plant.

The objective of this chapter is to develop the state of knowledge or the state-of-the-art of the technology of virus removal and inactivation during treatment of municipal wastewaters and water supplies. The effectiveness of municipal wastewater treatment will dictate to a large extent the concentration of viruses discharged into the aquatic system. Treatment of water supplies therefore provides the final barrier against viruses coming into contact with the population through drinking water.

The effective treatment of wastewaters to meet national goals of pollution elimination and providing safe drinking waters will result in greater quantities of sludges and other residuals that contain viruses. These virions in many cases can be eluted from the sludges. Therefore, selection of sludge disposal alternatives warrants consideration of potential problems with release of viruses. Detailed discussion of the disposal of virus-laden sludges is beyond the scope of this chapter.

VIRUS INACTIVATION DURING MUNICIPAL WASTEWATER TREATMENT

The effective treatment of municipal wastewater generally involves a sequence of processes, each designed to concentrate the undesirable component (pollution) in the wastewater into the smallest possible volume, and to release a liquid effluent of relatively good quality that meets effluent guidelines prescribed by regulatory agencies. The sequence and the combinations of the unit processes that can be used to produce an effluent of an acceptable quality are myriad. In general, the processes can be characterized as preliminary, primary, secondary (biological) and advanced treatment.

The actual virus concentration in untreated municipal wastewater varies considerably from country to country. In the United States, reported titers range from a low of about 200 plaque-forming units per liter (PFU/l) in cold weather to about 7000 PFU/l in warmer months (Clarke *et al.*, 1964; Kelly, 1960). On the other hand, in South Africa untreated municipal wastewaters were reported to contain in excess of 100,000 PFU/l (Nupen *et al.*, 1974).

The reported virus titers must be viewed with care. There is no universal procedure for the cultivation of all viruses. Each procedure is selective in respect to the viruses enumerated and is affected by the method of concentrating the viruses in the sample, selection of a host cell and the type of culture techniques used. In some procedures, the suspended solids in the sample are removed by filtration or centrifugation prior to virus concentration and enumeration. Viruses are associated with suspended solids (Moore *et al.,* 1975) and also may be embedded in human excrement. Therefore, by separation of the solids from the aqueous phase these viruses are not included in the total titer reported.

It is safe to say, however, that a relatively large quantity of viruses enter wastewater treatment plants daily. Virus survival during treatment of municipal wastewater has been studied in field-scale operations and even more intensely in the laboratory in bench-scale units. There is a wide variation in the concentration of indigenous enteroviruses and bacteriophages in untreated and partially treated wastewaters. The nonspecificity of virus enumeration techniques also leads to problems of data interpretation. Therefore, many of the reported data on virus removal were observed in full-scale and bench-scale systems, which were inoculated with model viruses to evaluate the inactivation and/or removal capabilities of various treatment processes. It should be pointed out here that in most of these studies the virus titers used were far in excess of the indigenous concentrations (4000-7000 PFU/l)common in wastewaters in the United States. Extrapolation of data observed for the bacteriophage to inactivation of enteric animal viruses cannot be done because in most cases the phages are not good models of the behavior of enteric animal viruses in the natural environment.

PRIMARY TREATMENT

The data presented in Table I indicate that removal of viruses during primary treatment ranged from almost zero to relatively high virus removal. Viruses are associated with suspended solids whether by adsorption or being embedded in the organic material. This association of virus with particulates coupled with the variability of the efficiencies of the different virus enumeration techniques may account for this wide discrepancy in the reported data.

Table I. Virus Removal During Primary Clarification

Model Virus	Titer (PFU/l)	Removal %	Reference
Bacteriophage F2	$(6.7\text{-}7.6) \times 10^5$	37.1	Sherman, 1975
Polio 1 (Mahoney)	2×10^8	26-55	Clarke *et al.,* 1964
Polio 1, 2, 3 (Sabin)	a	0-12	England *et al.,* 1967

[a]Natural level following immunization.

The removal of bacteriophage F2 is reported by Sherman (1975) for two municipal plants. Virus removal in the grit chamber-comminutor section of the plant ranged from 0-52.8% with an average of 18.4% at a flow rate of 2.5 million gallons per day (mgd). However, at the second plant, treating 1.5 mgd, the virus removal in this preliminary treatment unit ranged from 0-21.2% with an average of 7.1%.

The average bacteriophage F2 removal during primary treatment at a flow rate of 2.5 mgd was 37.1% with a range of 27.8%-54.2%. The observed data for the plant handling 1.5 mgd shows the virus removal ranging from 13.4%-69.4% in the primary basin with an average value of 32.2%.

Increases in the concentration of viruses in the primary settling tank effluent also have been reported at some treatment plants and may be the result of the release of viruses from the settled sludge. This observation is possible if the suspended solids in the influent are excluded from the virus enumeration technique.

BIOLOGICAL TREATMENT

During biological treatment, a portion of the oxygen-demanding dissolved organic material is converted into bacterial cell mass. Viruses may be entrapped in this biomass, which is in the form of a suspended growth in the activated sludge system and the various modifications of this process. Similar reactions take place in the trickling filter of the rotating bio-disc, in which the biological mass grows attached to a surface. The biomass developed in this biological reactor is separated from the liquid phase in a secondary clarifier, which is an integral part of these treatment systems.

Another biological treatment system commonly used to treat municipal wastewater is the waste stabilization pond, in which bacteria convert the oxygen-demanding dissolved organic material into cell mass, carbon dioxide and water. The cell mass settles to the bottom of the pond and undergoes decomposition. Algae also grow in the ponds and produce oxygen through photosynthetic reactions. The efficiencies of these three basic systems in the removal of viruses are discussed below.

ACTIVATED SLUDGE

The activated sludge system has shown to be efficient in virus inactivation (Table II). Continuous flow bench-scale activated sludge systems were operated by Ranganathan *et al.* (1975). The indigenous virus titer in the municipal wastewater used in these studies was consistently low and quite variable; therefore, poliovirus 1 (vaccine) was used as an inoculum. A 2-$\log_{10}$ decrease in the poliovirus concentration was observed in each of three reactors. The effluent poliovirus concentration ranged from 12 PFU/l-480 PFU/l. The poliovirus became associated with the suspended solids, and settling of the sludge solids accounted for almost a one-log inactivation. Poliovirus could be recovered from the sludge, and the data indicate recoveries of 13-5750 PFU/g of dry solids. Virus recovery from the solids increased as the organic loadings to the systems increased. However, a mass balance indicates permanent inactivation of a large percentage of the viruses added to the system. Therefore, the activated sludge has some inherent inactivating capabilities that are quite distinct from adsorption mechanisms attributed to other types of solids.

Table II. Virus Removal by the Activated Sludge Process

Model Virus	Titer (PFU/l)	Removal %	Reference
Bacteriophage T2	$(3\text{-}50) \times 10^5$	98	Kelly *et al.*, 1961
Coxsackie A9	3×10^8	96-99.4	Clarke *et al.*, 1961
Polio 1 (Sabin)	7.7×10^4	98	Malina *et al.*, 1974a
Polio 1 (MK 500)	$(2\text{-}200) \times 10^6$	64-78	Kelly *et al.*, 1961
Polio 1 (Mahoney)	7×10^7	79-94	Clarke *et al.*, 1961
Polio 1, 2, 3 (Sabin)	a	76-90	England *et al.*, 1967
Reovirus, Enterovirus	$<2 \times 10^5$ [b]	c	Malherbe and Strickland-Cholmley, 1967b

[a]Natural level following immunization.
[b]$TCID_{50}/1$ = Tissue Culture Infectious Dose.
[c]Qualitative data only.

The use of high-purity oxygen in laboratory-scale reactors instead of air for aerating activated sludge does not enhance virus removal (Ranganathan *et al.*, 1975). The data indicate that poliovirus removal efficiencies for the two systems range between 97.5 and 98.7% for aerated systems and between 97.1 and 99.3% for the oxygenated system. Dispersed growth was observed in the oxygenated reactor, which is possibly caused by shearing action of the mechanical mixing in the system. Therefore, if adsorption was the predominant virus removal mechanism, the observed virus removal with dispersed sludge would exceed that observed for well-flocculated sludge. However, enteric virus removal efficiencies for the two systems were the same. Therefore, other factors such as metabolic products or other characteristics of the biomass affect the virus inactivation.

The permanent inactivation of viruses in the activated sludge process was evaluated using Tritium-labeled poliovirus (Mahoney) added to activated sludge in batch reactors (Malina *et al.*, 1975). The radioactive label and the infectivity were traced with time. The labeled virus adsorbed to the solids in the mixed liquor almost immediately. After 10 minutes contact,

95% of the labeled virus was associated with the sludge solids. The radioactivity in the sludge remained unchanged for the next 15 hours of aeration, during which time 99.9% of the infectivity was removed from the supernatant. However, only 80% of the initial poliovirus infectivity was lost from the sludge during the 15-hour aeration period. Infective viruses were recovered from the suspended solids by shaking the sludge with deionized distilled water 15 minutes or dousing the sludge with triptose phosphate broth.

Enteric virus inactivation efficiency was monitored in a full-scale contact stabilization plant (Moore, 1974). The design flow of the plant was ten mgd. The data presented in Table III indicate that the contact stabilization process was effective in reducing the virus concentration 91-95%. At flows of approximately 7-8 mgd, the virus inactivation efficiency dropped to about 79%. However, in all four runs, the virus concentration in the plant effluent after chlorination was less than 10 PFU/l in all cases.

Table III. Virus Removal by the Contact Stabilization Process[a]

Design Flow 10 mgd				
	Virus Concentration			
		Mixed Liquor		
Flow (mgd)	Influent Wastewater (PFU/l)	Solids (PFU/l)	Supernatant (PFU/l)	Virus Removal Efficiency Contact Process Only (%)
≅ 7	1180	950	90	93
≅ 7	760	600	40	95
≅ 8	770	450	70	91
≅15	1490	900	310	79

A multistage pilot plant activated sludge system consisting of modified aeration, nitrification, denitrification with alum addition and filtration was operated by the Environmental Protection Agency in Washington, D.C. The overall efficiency of the plant and the removal of F2 bacteriophage was 99.97% (Safferman and Morris, 1976). However, the activated sludge component of the system accounted for an average removal of bacteriophage of 95%, but the range of removal efficiencies was 91.5-98.7%. Influent bacteriophage concentrations varied from 5.5×10^4 to 8×10^5 PFU/l.

Poliovirus inactivation in aerated lagoon systems operated at detention times of 1-10 days and at mixed liquor suspended solids concentrations of 238-370 mg/l produced effluents that were devoid of any detectable poliovirus (Ranganathan *et al.*, 1975). Influent concentrations of poliovirus ranged from 4.0 x 10^2 to 5.5 x 10^4 PFU/l. These systems proved equal or more effective in poliovirus removal than the activated sludge systems. The mechanisms of removal are still being postulated, but the long detention times are thought to be a major factor responsible for the high efficiency of poliovirus removal.

TRICKLING FILTER

The performance of trickling filters in virus removal is summarized in Table IV. Average virus removal in the trickling filter operating at a flow rate of 2.5 mgd was 18.9% although the removal efficiencies range from 6.3-40.4% (Sherman, 1975). The trickling filter at a flow rate of 1.5 mgd accounted for a range of virus removal of 1.53%-13.1% with an average of 9%. The above data are reported for the trickling filter alone and did not include the secondary sedimentation basin. The overall virus removal for the trickling filter and secondary clarifier for the 2.5 mgd plant increased to between 26.2 and 91.5% with an average of 49%. Similar results for the plant operating at a loading of 1.5 mgd ranged from 12.7-74.6%. The average removal efficiency was 37.4%.

Table IV. Virus Removal by Trickling Filters

Model Virus	Titer (PFU/l)	Removal %	Reference
Bacteriophage F2	(5.9-7.5) x 10^5	18.9	Sherman, 1975
Coxsackie A9	3 x 10^9	94	Clarke and Chang, 1975
Echovirus 12	7 x 10^9	83	Clarke and Chang, 1975
Polio 1	4 x 10^9	85	Clarke and Chang, 1975

Results of bench-scale studies using rotating tube trickling filters indicated similar overall inactivation efficiencies for trickling filters, namely 59% for poliovirus 1, 63% for echovirus 12, and 81% for Coxsackie A9 (Clarke *et al.*, 1961). These data indicate that the trickling filters are less efficient in the removal of viruses than the activated sludge process.

STABILIZATION PONDS

The removal of poliovirus in laboratory-scale oxidation ponds exceeded 99% under a wide range of loadings from 250 lb BOD_5/acre/day up to 550 lb BOD/acre/day (Malina and Melbard, 1974b). Other data, presented in Table V, also indicate the effectiveness of this process. The long detention time in ponds may account for the inactivation of the viruses. Viruses associated with influent suspended solids are removed when the solids settle to the bottom of the ponds. Attempts to recover viruses from solids that accumulated in model ponds were relatively unsuccessful (Moore *et al.*, 1975).

Table V. Virus Removal in Stabilization Ponds

Model Virus	Titer (PFU/l)	Removal %	Reference
Polio 1 (Sabin)	1.6×10^3	99	Ranganathan *et al.*, 1975
Polio 1 (attenuated)	5.6×10^5	92	Malina *et al.*, 1975
Polio 1 (Sabin)	3.3×10^3	99	Malina and Melbard, 1974b
Reovirus	20,000[a]	95	Nupen *et al.*, 1974
Polio 1 (Mahoney)	$(6\text{-}18000) \times 10^3$	99.97- None detected	Malina and Melbard, 1974b

[a] $TCID_{50}/l$ = Tissue Culture Infectious Dose.

DISINFECTION

The final unit treatment process of municipal wastewaters prior to discharge into the natural environment is disinfection. Traditionally, chlorine has been the major disinfecting compound used; however, the concern over the effects of chlorinated hydrocarbons in natural waters has drawn more attention to application of ozone for disinfection in the U.S. Ozone has been used for disinfection in Europe for both water supplies and effluents of municipal wastewater treatment plants.

CHLORINE

Viral resistance to chlorine varies from one type of virus to another. Required dosage of chlorine for inactivation of viruses is controlled by the chemical quality of the water; specifically, organic compounds, ammonia and suspended solids. These materials will react with the chlorine and reduce the amount of free chlorine available for virus inactivation. A summary of data relating to inactivation of virus by chlorine is presented in Table VI.

Table VI. Virus Inactivation by Chlorine

Model Virus	Titer (PFU/l)	Removal %	Reference
Bacteriophage F2	No data	99.997	*Water Sewage Works,* 1975
Polio	10^{9}[a]	99	Kott *et al.*, 1975
Bacteriophage F2	No data	99	Longley *et al.*, 1974

[a] $TCID_{50}/l$ = Tissue Culture Infectious Dose.

There seems to be a lack of unanimity among various investigators regarding the optimum pH for virus inactivation. Berg (1973) reported that 0.1 mg/l of HOCl destroyed 99% of the viruses in surface and renovated waters after 100 minutes of contact time. Therefore, the pH conducive for virus inactivation would be between pH 4 and pH 7 since chlorine exists in water predominantly as HOCl in this pH range. Results with poliovirus by Kott *et al.* (1975) support inactivation in this range of pH. A 99% decrease of poliovirus in distilled deionized unbuffered water was achieved for a chlorine concentration of 0.4 mg/l at a pH 6 after 28 seconds of contact time. However, at pH 10 the required time was 107 seconds. As the chlorine concentration was increased to 0.8 mg/l in the same water system, 99% decrease of poliovirus was achieved at pH 6 after 16 seconds of contact time, while 42 seconds were required at a pH 10. The interference of materials in reclaimed water was readily recognized since 99% decrease in poliovirus was achieved at a chlorine concentration of 0.4 mg/l at pH 6 only after 46 seconds of contact. The required time at the 0.4 mg/l concentration at a pH 10 in reclaimed water was 166 seconds. Increasing the dose to 0.8 mg/l reduced the contact time to 22 seconds at a pH 6. However, 168 seconds were required for 99% decrease in poliovirus at a pH 10.

No detectable viruses were observed in the chlorinated effluent in the South Lake Tahoe Water Reclamation Plant (Culp, 1974). The data were observed over a four-month period and indicate the consistency of chlorination in inactivating viruses at that location.

The type of chlorine compound used and the degree of turbulence at the point of chlorine addition also seem to affect the dose of chlorine required and the inactivation of virus achieved. At the same dose chlorine gas, when applied under controlled mixing conditions to wastewater, provided better disinfection than addition of aqueous chlorine (Longley *et al.*, 1974). The data reported indicated excellent viral inactivation almost instantaneously using gaseous chlorine at approximately pH 3 of the wastewater. In these studies the F2 virus was the model virus. Introduction of gaseous chlorine into a double set of nozzles resulted in the removal of more than 99.9% of F2 viruses in less than 100 seconds contact time in wastewaters (*Water Sewage Works,* 1975). Using aqueous chlorine, 99% of the F2 viruses were inactivated in approximately 100 seconds contact time.

Cramer *et al.* (1976) evaluated the resistance of polio 3 and bacteriophage F2 to chlorine. Chlorine dose of 30 mg/l was introduced to water at various pH values prior to the introduction of virus. The chlorine residuals decreased as the pH increased, and at these reduced residual chlorine concentrations the times required for 90% removal of poliovirus were 3, 5 and 15 minutes, respectively, for the systems at pH 6, pH 7 and pH 10.

The second set of conditions involved the introduction of the chlorine in the system in which poliovirus 3 already were present. The required time for 90% virus inactivation was 0.17 minutes at all three pH values. The observed data also indicate that virus inactivation was retarded at pH values below pH 4.

Iodine is an effective virucidal agent (Table VII). Doses of 10.2 mg/l at pH 10.0 resulted in 99.999% reduction in viruses after only one minute contact time. Kruse *et al.* (1971) indicated that the combined forms of halogens have very little virucidal effect. Attempts to improve inactivation by step dosage of the halogens proved to be of no consequence.

Table VII. Virus Inactivation by Iodine

Model Virus	Titer (PFU/l)	Removal %	Reference
Bacteriophage F2	10^9	99.9999	Cramer *et al.*, 1976
Polio 3	No data	99.9999	Cramer *et al.*, 1976

BROMINE

The use of bromine for disinfection shows promise for the future (Floyd *et al.*, 1976). A loss of one $\log_{10}$ for 10-second contact time at pH 7 was reported at a temperature of 2°C. Much higher inactivation rates were observed at 10°C and 20°C.

OZONE

The data presented in Table VIII indicate the high efficiency and consistency of ozone as a disinfectant. Ozone does not result in formation of any chlorinated organic compounds and offers an effective way of disinfecting effluents or water supplies.

Table VIII. Virus Inactivation by Ozone

Model Virus	Titer (PFU/l)	Removal %	Reference
Bacteriophage F2	1×10^{11}	100	Pavoni and Tittlebaum, 1974
Coxsackie B3	2.5×10^2	99.9	Keller, 1974
Polio 1	$(1.4\text{-}6.3) \times 10^7$	99.994	Majumdar *et al.*, 1974
Polio 2	5×10^2	99.99	Keller, 1974

Katzenelson (1974) reported that poliovirus 1 and bacteriophage F2 were inactivated at ozone concentrations of 0.01-2.5 mg/l of ozone at much shorter contact times than required with similar concentrations of chlorine.

Almost complete inactivation of bacteriophage F2 was achieved at a concentration of ozone of 15 mg/l (Pavoni and Tittlebaum, 1974). In these studies, the rate of viral inactivation exceeded the rate of bacterial inactivation. Essentially 100% inactivation of viruses in secondary treatment plant effluent was reported after five minutes contact time and ozone residuals of 0.15 mg/l. The inactivation of poliovirus in a continuous flow system is demonstrated by Majumdar *et al.* (1974). The average virus inactivation increased with increasing ozone concentration and required lower contact times. In treating primary effluents, concentration of 0.84 mg/l of ozone at a contact time of eight minutes resulted in greater than 98% inactivation of the poliovirus. Increasing the concentration to 4.4 mg/l and reducing the residence time to one minute increased the virus inactivation to about 99.99%. Similar results were observed in treating secondary treatment plant effluents. A threshold concentration for virus inactivation was considered to be 1.0 mg/l. Keller (1974) used ozone and achieved 99.9% inactivation of Coxsackie B3 and polio 2. These data observed during pilot plant studies indicate that removals as high as 99.999% of Coxsackie B3 were possible.

VIRUS INACTIVATION DURING TREATMENT OF WATER SUPPLIES

Treatment of surface water to provide municipal drinking water supplies usually involves chemical precipitation and/or coagulation and flocculation followed by sedimentation, filtration and disinfection. At times, activated carbon is added to mitigate problems with tastes and odors. The very same processes, for the most part, also are used for the advanced treatment of wastewaters. Therefore, the following discussion is directed at virus removal by the chemical and physical processes used in the treatment of water, whether the source is a lake or stream or biological waste treatment plant effluent, since many of the problems are similar, and the virus removal efficiencies should be identical.

COAGULATION

A summary of reported data on the effectiveness of the coagulation process in removing viruses from water is presented in Table IX.

The chemical characteristics affect the quantity of chemical coagulant required and the efficiency of the process. Aluminum sulfate and ferric chloride were reported to be effective coagulants for the removal of physical and chemical constituents in water as well as for the removal of bacteriophage F2 (York and Drewry, 1974). In excess of 99% of the initial virus titer and 90% of the turbidity were removed during coagulation with alum at a dose ranging between 20 and 24 mg/l while approximately 50 mg/l of ferric chloride was required. Anionic, cationic and nonionic polyelectrolytes, when used as coagulant aids with alum, did not improve the removal of bacteriophage F2, but did enhance floc formation. Two cationic polyelectrolytes used as coagulants did remove 76% and 99% of bacteriophage F2, respectively, but did not remove turbidity and exhibited poor floc formation.

Table IX. Virus Removal by Coagulation

Model Virus	Titer (PFU/l)	Removal %	Reference
Bacteriophage	$2\text{-}6 \times 10^5$	99	Chang, 1958b
Bacteriophage F2	1×10^9	76-99.8	York and Drewry, 1974
Bacteriophage F2	2.5×10^6	>99.7	Wolf *et al.*, 1974
Bacteriophage MS2	$2\text{-}5 \times 10^8$	99.8	Chaudhuri, 1974
Bacteriophage MS2	$3\text{-}4 \times 10^8$	99.7	Manwaring *et al.*, 1971
Bacteriophage T2	$3\text{-}7 \times 10^7$	94	Thorup *et al.*, 1970
Bacteriophage T4	1×10^7	83	Gilcreas and Kelly, 1955
Coxsackie A2	1×10^6	99	Chang, 1958b
Polio 1 (Sabin)	$10^8\text{-}10^{10}$	86	Thorup *et al.*, 1970
Polio 1	1.13×10^5	>99.7	Wolf *et al.*, 1974

The results of virus inactivation during physical chemical treatment of a biological treatment plant effluent in a large pilot plant were reported by Wolf *et al.* (1974). The removal of bacteriophage F2 was 99.845% for coagulation with alum followed by sedimentation, and the removal increased to 99.985% when the filtration process was added to the coagulation-sedimentation system. These virus removal efficiencies were observed at an Al:P ratio of 7:1, and the efficiencies decreased markedly at lower alum doses. Bacteriophage F2 removals dropped to only 46% at an Al:P ratio of 0.44:1 in the coagulation-sedimentation system, while the poliovirus removal efficiency was only 63% under the same conditions.

The effect of high pH ($pH > 11.0$) on virus removal also was demonstrated by Wolf *et al.* (1974). No viable poliovirus 1 particles were recoverable from sludge or effluent of a high lime treatment unit. These results must be interpreted with caution since the virus inoculum was added to the liquid phase and the viruses were not enmeshed in the indigenous suspended solids in the effluent of the biological treatment plant. Therefore, the viruses were not protected from the adverse high pH environment by enmeshment in the solids. The effectiveness of high pH on virus inactivation also was reported by Nupen *et al.* (1974), who observed a four-to-five $\log_{10}$ reduction in poliovirus 2 titer ($TCID_{50}$) during coagulation with lime at between pH 11.3 and pH 11.5 and a detention time of 60 minutes. Virus inactivation increased with increase in pH and time of exposure. Virus could be recovered from a one-liter sample of effluent after exposure to pH 11.5 for three hours. However, no virus was recovered from 10 liters of settled lime floc.

The removal of viruses during water softening precipitation was investigated by Thayer and Sproul (1966). Bacteriophage T2 was inactivated to varying degrees depending on the pH and the type and quantity of precipitate formed. The bacteriophage was essentially completely inactivated at pH 10.8. Excess lime treatment resulted in 99.95% phage removal.

FILTRATION

Filtration using sand, anthracite and combinations of these two granular media, or using diatomaceous earth, commonly is used to remove suspended solids that pass through clarifiers following the coagulation process. Sand and anthracite virus removal by sand filters have been reported and are summarized in Table X. Robeck *et al.* (1962) reported more than 98% removal of poliovirus 1 during filtration at rates of 2 and 6 gpm/ft^2 through 16 inches of coarse anthracite on top of 8 inches of sand. Filter runs were increased when small doses of alum were well mixed in water prior to filtration; however, the virus removal efficiency remained the same. When treating turbid waters virus penetration accompanied floc breakthrough, and the addition of 0.05 mg/l of polyelectrolyte helped prevent breakthrough. In general, virus penetration increased with effluent turbidity, especially at a low alum concentration, namely 5 mg/l.

Table X. Virus Removal by Sand Filtration

Model Virus	Titer (PFU/l)	Removal %	Reference
Bacteriophage T4	1×10^7	99	Gilcreas and Kelly, 1955
Polio 1 (Mahoney)	$1\text{-}6 \times 10^4$	>99	Robeck *et al.*, 1962
Reovirus, Enterovirus	$<2 \times 10^{5}$ [a]	b	Malherbe and Strickland-Cholmley, 1967a

[a]$TCID_{50}$/l = Tissue Culture Infectious Dose.
[b]Qualitative data only.

DIATOMACEOUS EARTH

Virus removal by the diatomaceous earth filtration process has been evaluated using different grades of untreated diatomaceous earth, and organic as well as inorganic coatings on the filter aids (Table XI).

Table XI. Virus Removal by Diatomaceous Earth Filtration

Model Virus	Titer (PFU/l)	Removal %	Reference
Bacteriophage T2	(1.44-3.48) x 10^3	100	Brown *et al.*, 1974a
Bacteriophage T4	3-4 x 10^8	99	Chaudhuri *et al.*, 1974
Bacteriophage MS2	3-4 x 10^8	99	Chaudhuri *et al.*, 1974
Bacteriophage MS2	1.2 x 10^8	99.999	Amirhor and Engelbrecht, 1975
Polio 1 (Mahoney)	3.48 x 10^3	99	Brown *et al.*, 1974b

Uncoated diatomite (Hyflo) initially removed more than 90% of the influent virus, followed by slowly diminishing efficiencies (Brown *et al.*, 1974a, 1974b). The capacity of the uncoated diatomite to remove poliovirus was 62% compared to 90% for bacteriophage T2 after two hours filtration. Greater than 98% removals of the bacteriophage T2 and poliovirus were observed using Hyflo coated with either ferric hydrate or C-31 polymer. However, the procedure used to coat the diatomite can significantly influence the virus removal efficiency. Pretreatment of the influent water with C-31 polymer and subsequent filtration with Hyflo resulted in reduction of the virus titers to below the detectable limits of the virus enumeration techniques used.

ADSORPTION ON ACTIVATED CARBON

Activated carbon is used to remove tastes and odors from water supplies and refractory materials from wastewaters. Viruses also are adsorbed to carbon to various degress (Table XII). Gerba *et al.* (1974) reported that the adsorption of polioviruses to activated carbon is defined by the Freundlich isotherm. The adsorption was dependent on pH, and the maximum removals were observed at pH 4.5. Adsorption was reported to be minimal at pH 7.5 to pH 8.5. Sproul (1968) reported that the adsorption of bacteriophage T2 varied from 29-75% depending on the type of carbon.

Table XII. Adsorption of Viruses of Activated Carbon

Model Virus	Titer (PFU/l)	Removal %	Reference
Bacteriophage T2	6-9 x 10^8	75	Sproul, 1968
Bacteriophage T4	1.44 x 10^7	70	Oza and Chaudhuri, 1975
Bacteriophage T4	1 x 10^{11}	99	Cookson, 1967
Polio 1	8.8 x 10^6	90	Gerba, 1974

Maximum adsorption of bacteriophage T4 on coal was observed at pH 8.0 and on ionic strength of 0.015 (Oza and Chaudhuri, 1975). They reported that adsorption of bacteriophage T4 on coal was described by the Langmuir isotherm. The information relating to adsorption of viruses on activated carbon is contradictory; however, activated carbon seldom is used as the only process to produce a finished water. Virus removal and inactivation can be accomplished with a higher degree of certainty, by other processes preceding (coagulation-sedimentation-filtration, or following disinfection) the adsorption unit.

REVERSE OSMOSIS

Reverse osmosis and other membrane processes are used more frequently in advanced wastewater treatment schemes than to provide drinking water supplies. A limited amount of work has been reported on virus rejection by reverse osmosis (Table XIII).

Table XIII. Virus Removal by Reverse Osmosis

Model Virus	Titer (PFU/l)	Removal %	Reference
Bacteriophage T2	6.8×10^8	$\geqslant 99$	Sorber *et al.*, 1971
Poliovirus 1	5.5×10^1	$\geqslant 99$	Sorber *et al.*, 1971

The data observed by Sorber *et al.* (1971) indicate that the virus concentration in the product water was relatively constant, namely 0-7.6 PFU/l for bacteriophage T2 and 0-0.71 PFU/l for poliovirus. These data support the theory of random virus penetration of cellulose acetate membranes, especially at feed water virus concentrations below 10^8 PFU/l. Specifically, at naturally occurring virus levels, viruses in the feed water to a reverse osmosis or ultrafiltration process employing asymmetrical cellulose membranes can randomly penetrate imperfections in the membranes.

From a practical standpoint, cellulose acetate membranes cannot be expected to provide a virus-free product water since virus penetration is random. Disinfection of the product water from reverse osmosis and ultrafiltration membranes is essential to the production of virus-free water for drinking supplies. It should be noted, however, that product disinfection would be more efficient after reverse osmosis or ultrafiltration since considerable quantities of particulate and dissolved materials exerting a disinfectant demand would have been eliminated from the product water of these unit processes.

SUMMARY

Municipal wastewater treatment plants incorporating primary sedimentation, biological treatment and disinfection can produce effluents that contain less than 10 PFU/l of enteric animal viruses. Adding advanced treatment to process this effluent (Figure 1) should produce a water in which no viruses can be detected, based on current techniques available for virus concentration and enumeration.

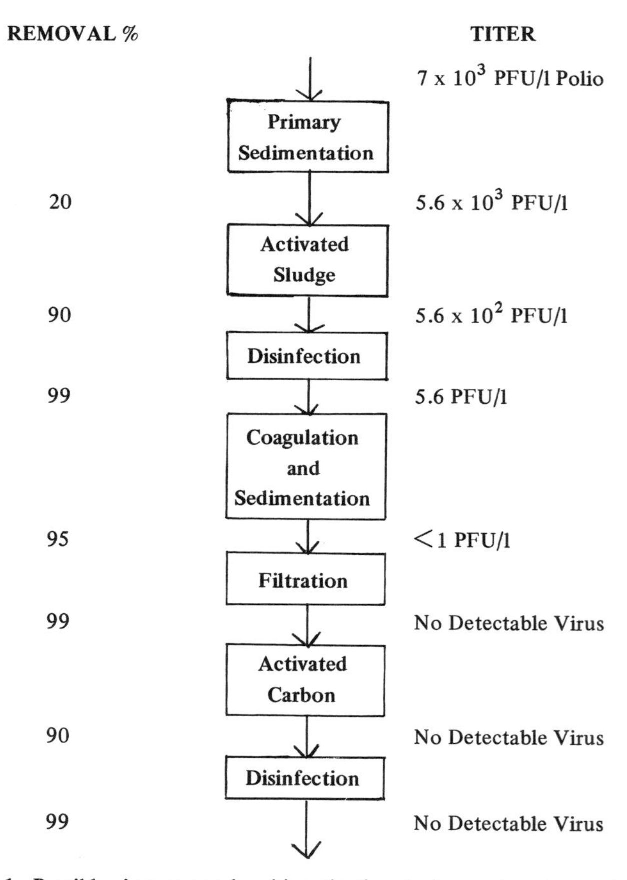

Figure 1. Possible virus removal and inactivation during wastewater treatment

It is impossible to say that there will be no virions in the effluent of the proposed system when operating under actual field conditions. However, the product water is essentially virus-free. The lack of transmission of viral disease via drinking water supplies also provides evidence of the effectiveness of chemical coagulation, filtration and disinfection in the removal and inactivation of viruses found in surface waters.

REFERENCES

Amirhor, P. and R. S. Engelbrecht. "Virus Removal by Polyelectrolyte Aided Filtration," *J. Am. Water Works Assoc.* 67(4):187 (1975).

Berg , G. "Reassessment of the Virus Problem in Surface and Renovated Waters," in *Water Quality Management and Pollution Control Problems,* S. Jenkins, Ed. (London and Aylesbury, England: Compton Printing Ltd., 1973), p. 92.

Berg, G. "Detection, Occurrence, and Removal of Viruses," *J. Water Poll. Control Fed.* 47(6):1587 (1975).

Brown, T. S., J. F. Malina, Jr. and B. E. Moore. "Virus Removal by Diatomaceous Earth Filtration–Part I," *J. Am. Water Works Assoc.* 66:98 (1974a).

Brown, T. S., J. F. Malina, Jr. and B. E. Moore. "Virus Removal by Diatomaceous Earth Filtration–Part II," *J. Am. Water Works Assoc.* 66:735 (1974b).

Chang, S. "Removal of Coxsackie and Bacterial Virus in Water By Flocculation–Part I," *Am. J. Public Health* 48:51 (1958a).

Chang, S. "Removal of Coxsackie and Bacterial Virus in Water by Flocculation–Part II," *Am. J. Public Health* 48:159 (1958b).

Chaudhuri, M., P. Amirhor and R. S. Engelbrecht. "Virus Removal by Diatomaceous Earth Filtration," *Am. Soc. Civil Eng. Environ.* 100:937 (1974).

Clarke, N., R. E. Stevenson, S. L. Chang and P. W. Kabler. "Removal of Enteric Viruses from Sewage by Activated Sludge," *Am. J. Public Health* 51:1118 (1961).

Clarke, N., G. Berg, P. W. Kabler and S. L. Chang. "Human Enteric Viruses in Water: Source, Survival and Removability," *Advances in Water Pollution Research, Proceedings International Conference,* London 2 (London: Pergamon Press, 1964).

Clarke, N. and S. Chang. "Removal of Enteroviruses from Sewage by Bench-Scale Rotary-Tube Trickling Filters," *Appl. Microbiol.* 30:233 (1975).

Cookson, J. "Adsorption of Viruses on Activated Carbon: Equilibria and Kinetics of the Attachment of E. Coli-Bacteriophage T4 on Activated Carbon," *Environ. Sci. Technol.* 1:46 (1967).

Cramer, W., K. Kawata and C. W. Kruse. "Chlorination and Iodination of Poliovirus and F2," *J. Water Poll. Control Fed.* 48:61 (1976).

Culp, R. "Breakpoint Chlorination for Virus Inactivation," *J. Am. Water Works Assoc.* 66:699 (1974).

England, B., R. E. Leach, B. Adams and R. Shiosak. "Virologic Assessment of Sewage Treatment at Santee California," in *Transmission of Viruses by the Water Route,* G. Berg, Ed. (New York: Wiley-Interscience, 1967).

Floyd, R., J. D. Johnson and D. G. Sharp. "Inactivation by Bromine of Single Poliovirus Particles in Water," *Appl. Microbiol.* 31:298 (1976).
Foliguet, J. and F. Doncoeur. "Removal of Viruses from Water by Breakpoint Chlorination," *Water Res.* 8:651 (1974).
Foliguet, J. and F. Doncoeur. "Elimination des Enteroviruses au cours Traitement des eaux d'alimentation par Coagulation-Floculation-Filtration," *Water Res.* 9:953 (1975).
Gerba, C. "Enhancement of Poliovirus Adsorption in Wastewater onto Activated Carbon," in *Virus Survival in Water and Wastewater Systems,* J. F. Malina, Jr. and B. P. Sagik, Eds. Center for Research in Water Resources, The University of Texas, Austin, Texas (1974), p. 115.
Gilcreas, F. and S. Kelly. "Relation of Coliform Organism Test to Enteric Virus Pollution," *J. Am. Water Works Assoc.* 47:683 (1955).
Katzenelson, E."Inactivation of Virus and Bacteria by Ozone," *Chem. Abstr.* 81:281. (1974).
Keller, J. "Ozone Disinfection Pilot Plant Studies at Laconia, New Hampshire," *J. Am. Water Works Assoc.* 66:734 (1974).
Kelly, S. M. and W. W. Sanderson. "The Density of Enteroviruses in Sewage," *J. Water Poll. Control Fed.* 32:1269 (1960).
Kelly, S. M., W. W. Sanderson and C. Neidl. "Removal of Enteroviruses from Sewage by Activated Sludge," *J. Water Poll. Control Fed.* 33: 1056 (1961).
Kott, Y., E. M. Nupen and W. R. Ross. "The Effect of pH on the Efficiency of Chlorine Disinfection and Virus Enumeration," *Water Res.* 9:869 (1975).
Kruse, C. W., V. Olivieri and K. Kawata. "The Enhancement of Viral Inactivation by Halogens," *Water Sew. Works* 118:187 (1971).
Longley, K. E., V. P. Olivieri, C. W. Kruse and K. Kawata. "Enhancement of Terminal Disinfection of a Wastewater System," in *Virus Survival in Water and Wastewater Systems,* Center for Research in Water Resources, The University of Texas, Austin, Texas (1974), p. 166.
Lund, E., D. E. Hedstrum, and N. Jantzen. "Occurrence of Enteric Viruses in Wastewater After Activated Sludge Treatment," *J. Water Poll. Control Fed.*41:169 (1969).
Majumdar, S. B., W. H. Ceckler and O. J. Sproul. "Inactivation of Poliovirus in Water by Ozonation," *J. Water Poll. Control Fed.* 46:2048 (1974).
Malherbe, H. and M. Strickland-Cholmley. "Quantitative Studies on Survival in Sewage Purification Processes," in *Transmission of Viruses by the Water Route,* G. Berg, Ed. (New York: Wiley-Interscience 1967a).
Malherbe, H. and M. Strickland-Cholmley. "Survival of Viruses in the Presence of Algae," in *Transmission of Viruses by the Water Route,* G. Berg, Ed. (New York: Wiley-Interscience, 1967b).
Malina, J. F., Jr., K. Ranganathan, B. P. Sagik and B. E. Moore. "Poliovirus Inactivation by Activated Sludge," in *Virus Survival in Water and Wastewater Systems,* Center for Research in Water Resources, The University of Texas, Austin, Texas (1974a), p. 95.
Malina, J. F., Jr. and A. Melbard. "Inactivation of Virus in Bench-Scale Oxygenated Waste Stabilization Ponds," Technical Report CRWR-109, EHE 74-02, The University of Texas, Austin, Texas (1974b).
Malina, J. F., Jr., K. Ranganathan, B. P. Sagik and B. E. Moore. "Poliovirus Inactivation by Activated Sludge," *J. Water Poll. Control Fed.* 47:2178 (1975).

Manwaring, J. F., M. Chaudhuri and R. S. Engelbrecht. "Removal of Viruses by Coagulation and Flocculation," *J. Am. Water Works Assoc.* 63:298 (1971).

Moore, B. E., B. P. Sagik and J. F. Malina, Jr. "Viral Association with Suspended Solids," *Water Res.* 9:197 (1975).

Moore, B. E. "Application of Viral Concentration Techniques to Field Sampling at an Activated Sludge Treatment Plant," Master's Thesis, The University of Texas, Austin, Texas (December 1974).

Nupen, E. M., B. W. Batement and N. C. McKenny. "The Reduction of Virus by the Various Unit Processes Used in the Reclamation of Sewage to Potable Waters," in *Virus Survival in Water and Wastewater Systems,* J. F. Malina, Jr. and B. P. Sagik, Eds. Center for Research in Water Resources, The University of Texas, Austin, Texas (1974).

Oza, P. and M. Chaudhuri. "Removal of Viruses from Water by Sorption to Coal," *Water Res.* 9:707 (1975).

Pavoni, J. and M. Tittlebaum. "Virus Inactivation in Secondary Wastewater Treatment Plant Effluent Using Ozone," in *Virus Survival in Water and Wastewater Systems,* J. F. Malina, Jr. and B. P. Sagik, Eds. Center for Research in Water Resources, The University of Texas, Austin, Texas (1974).

Ranganathan, K., J. F. Malina, Jr. and B. P. Sagik. "Inactivation of Enteric Virus During Biological Wastewater Treatment," in *Progress in Water Technology,* 7 (3/4), (London: Pergamon Press, 1975).

Robeck, G., N. A. Clarke and K. A. Dostal. "Effectiveness of Water Treatment Processes in Virus Removal," *J. Am. Water Works Assoc.* 54:1275 (1962).

Safferman, R. and M. Morris. "Assessment of Virus Removal by a Multi-Stage Activated Sludge Process," *Water Res.* 10:413 (1976).

Sherman, V. "Virus Removal in Trickling Filter Plants," *Water Sew. Works* R-36 (April 30, 1975).

Sorber, C. A., J. F. Malina, Jr. and B. P. Sagik. "Virus Rejection by the Reverse Osmosis-Ultrafiltration Processes," Technical Report EHE 71-9, CRWR-82, Center for Research in Water Resources, The University of Texas, Austin, Texas (1971).

Sproul, O. J. "Virus Removal by Adsorption in Treatment Processes," *Water Res.* 2:74 (1968).

Thayer, S. E. and O. J. Sproul. "Virus Inactivation in Water-Softening Precipitation Processes," *J. Am. Water Works Assoc.* 58:1063 (1966).

Thorup, R. T., F. P. Nixon, D. F. Wentworth and O. J. Sproul. "Virus Removal by Coagulation with Polyelectrolytes," *J. Am. Water Works Assoc.* 62:97 (1970).

Water Sewage Works. "New Chlorine Application Improves Viral Kills," R-68, (April 30, 1975).

Wolf, H. W., R. S. Safferman and A. R. Mixson. "Virus Inactivation During Tertiary Treatment," in *Virus Survival in Water and Wastewater Systems,* J. F. Malina, Jr. and B. P. Sagik, Eds. Center for Research in Water Resources, The University of Texas, Austin, Texas (1974).

York, D. and W. Drewry. "Virus Removal by Chemical Coagulation," *J. Am. Water Works Assoc.* 66:711 (1974).

SECTION II

TRACE ORGANICS

4

TRACE ORGANICS IN WATER AND WASTEWATER

Raymond P. Canale

Associate Professor
Department of Civil Engineering
University of Michigan
Ann Arbor, Michigan

INTRODUCTION

This chapter will introduce the general subject and problems of trace organic contaminants in water and wastewater and briefly describe methodologies that might ultimately be used to determine the spatial and temporal distribution of these materials in the environment. The presentation will include a discussion of the sources of both natural and man-made organics in natural waters. Persistent organics–those organics not readily subject to natural degradation processes–will be classified into major groups based on their chemical structure. Many of these organic compounds have important economic and health benefits to man and, therefore, their use is expected to increase in the future. Unfortunately, many of these same organic compounds have also been shown to be harmful to man and wildlife, and many others are suspect. It is the responsibility of engineers, toxicologists, public health officials and water resource and environmental decision-makers to minimize the potential hazards of these chemicals to man and other biota in the environment. These issues have been brought into sharp focus by recent EPA proposals dealing with the control of organic chemical contaminants in drinking water. Enactment of these proposals could require major changes in current water treatment practice. The resolution of these complex problems will require answers to many questions concerning the sources, health hazards, detection, monitoring and treatment of organics in water supplies.

In addition, other sources of harmful organics, such as those in our food supply, could have substantial impacts on health. Thus, it is also important to determine the significance of exposure from drinking water relative to other sources of organic compounds. A possible approach to the problem of determining the distribution of organics in the environment would be to utilize systems analysis and mathematical modeling techniques. Such techniques will be discussed using a long-term world model for DDT as an example.

SOURCES OF ORGANICS IN NATURAL WATERS

Organic contaminants are found in all natural waters and water supplies. These contaminants originate from diverse sources. Dissolved and particulate organic materials in water are natural phenomena associated with the life cycles of aquatic and terrestrial plants and animals. These materials are released continuously into the aquatic environment as algal and fungal excretions and metabolic by-products, as unassimilated organic material due to zooplankton grazing, and as a consequence of bacterial action on dead plants and animals. A large variety of organic compounds can be released, some of which are listed in Table I.

Table I. Sources of Organics in Natural Waters

Sources of Organics in Natural Waters
1. Decay of Plant and Animal Life
Humic and fulvic acids
2. Metabolic Activities and Excretions
Geosmin
2-Methylisoborneol
n-Heptanal
6-Pentyl-α Pyrone
Dimethyl sulfide
Isopropyl mercaptan
3. Industrial Wastes
Two million new compounds registered between 1965-1972
4. Domestic Wastes
Proteins
Carbohydrates
Fatty acid, etc.
30% unknown
New compounds produced during treatment
5. Nonpoint Sources
Agricultural–insecticides, herbicides
Urban–oils, solvents, fuels
Landfills–PCB
Atmospheric–insecticides

According to an AWWA committee on organics (1974), two million new industrial chemicals were registered by *Chemical Abstracts* between 1965 and 1972. Most were organics. This trend is expected to increase as our society becomes more complex. Organics enter water supplies as untreated effluents from industrial as well as domestic wastewater treatment plants. Several studies have been conducted to characterize the organic composition of secondary treated effluents (Painter *et al.,* 1961; Bunch *et al.,* 1961; Rebhum and Manka, 1974) and one study by Rosen *et al.,* (1972) has shown that additional compounds are synthesized during the secondary treatment process. Techniques employed to detect organics involve concentration and separation methods that isolate only fractions of the total organic content. Thus, all the studies involved unidentified portions of the original sample, which contain unknown organic compounds.

Organic contaminants enter natural waters from nonpoint sources, such as agricultural and urban areas, solid waste disposal sites, and in rain- and snowfall. Important organics that originate from nonpoint sources include insecticides, herbicides, fuels, oils, solvents and insulation fluids. Table I contains a summary of these compounds and their associated sources.

PERSISTENT ORGANIC CONTAMINANTS

Most of the sources of organic contaminants listed in Table I contain persistent fractions that are relatively resistant to physical or metabolic degradation. This type of organic has the greatest potential for harmful effects on man and the environment. As a matter of fact, some substances such as DDT have been developed and used specifically because of their chemical stability and toxic properties. Table II contains a list of groups of persistent organic chemicals as compiled by Holden (1975). Many of the chemicals listed have not been reported in the environment because of a lack of suitable techniques for detection.

Table II. Persistent Organic Chemicals

1. DDT, DDE, DDD, dieldrin, hexachlorobenzene
2. Polychlorinated biphenyls (PCB) and terphenyls (PCT)
3. Polychlorinated naphthalenes (PCN)
4. Chlorinated aliphatic hydrocarbons (chlorinated paraffins and olefins)
5. Halogenated solvents, certain refrigerants, fire retardants
6. Stable herbicides (paraquat, simazine)
7. Phthalic acid esters (plasticizers)
8. Straight-chain surfactants
9. Petroleum hydrocarbons
10. Polynuclear aromatics (benzophrenes)
11. Chlorinated dibenzodioxins and dibenzofurnas
12. Organometallic compounds (of mercury and tin)

BENEFICIAL USES OF ORGANIC CHEMICALS

Potentially harmful persistent organic chemicals are associated with almost every aspect of modern life, and this has been the situation for many years. That many of these chemicals are important to our health and economy ensure their continued and probably increased use.

As an example, malaria, typhus, plague and yellow fever have been largely controlled in many parts of the world by DDT since its first widespread use in 1942 (Hayes, 1975). It is estimated that about 950 million people now live in protected areas previously infested with malaria. Unfortunately, an equal number of people live in regions where the transmission of malaria still continues. Thus, on a worldwide basis, we can expect to see the continued use of such insecticides. With the control of such diseases several secondary economic benefits are derived, such as decreased hospitalization costs, reduced absenteeism from work, increases in land values and general enhancement of business and industrial activity.

Another example of the beneficial use of organic chemicals relates to the use of pesticides to help increase agricultural production through control of insects and excessive weeds. It has been estimated by Headley and Kneese (1969) that in the United States in 1963 the use of pesticides increased the value of farm products by about $1.8 billion. This increased production was derived from expenditures of about $.44 billion for pesticides. Such production has enabled the U.S. to maintain its food supply despite the fact that our population has grown steadily while the total area of land in production has declined.

Several other benefits are derived from use of pesticides. For example, wood products in the form of utility poles, railroad ties and dock pilings can be protected by impregnation with insecticides. It has been estimated by Hayes (1975) that the use of such chemicals saves as much wood in a single year as could be produced annually by a forest the size of New Hampshire. Other beneficial uses of herbicides include control of weeds on highways and utility right-of-ways.

Similar cost benefit analysis should be performed for other potentially harmful organic chemicals on a case-by-case basis. The continued use of a given chemical can only be justified on the basis of such an analysis along with a detailed evaluation of the potentially harmful impact on human health and wildlife.

HARMFUL EFFECTS OF ORGANIC CHEMICALS

Persistent organic chemicals can have harmful effects on man and other organisms in the environment. Such chemicals can be acutely toxic and cause severe illness or death following direct consumption of a sufficient dose. As an example, in 1968 hundreds of Japanese consumed rice oil contaminated with PCB (Ahmed, 1976). Severe cases of chloracne developed accompanied by vision impairment and neurological disorders. Recovery from the symptoms took several months. Continuous long-term, low-level intake of toxic organic materials through the water or food supply, or by direct contact can be the cause of chronic difficulties. The effects of such exposures in man may not be evident for several generations. In addition, two or more toxic organic chemicals may act together synergistically. Bioassay tests to assess the toxic effects of most organic contaminates have not been performed. However, numerous studies have shown that many compounds found in natural waters are carcinogenic in animals and potentially carcinogenic in man.

Organochloride pesticides have been the most widely studied type of organic contaminant in the environment. Because of their persistent nature and capacity for being transported long distances far from the point of application, such chemicals can have widespread impacts all over the globe, including in arctic areas.

Some persistent chemicals such as organochloride pesticides and PCB are much more soluble in fat than in water. Therefore, these chemicals will become concentrated through food webs in increasing concentrations to the top predator levels. The harmful effects of DDT on predatory birds is a well-known example. Typical concentration factors for DDT and PCB for various aquatic organisms are listed in Table III from data summarized by a panel on hazardous trace substances (1972).

Table III. Typical Concentration Factors for DDT and PCB

Organism	Conc. Organism/Conc. Water
Algae	700
Protozoa	50
Zooplankton	2000
Shrimp	100-200,000
Crab	5000
Oyster	5000
Panfish	700
Catfish	75,000
Bluegill	13,000
Salmon	60,000
Cod	200,000

RECENT HISTORY OF THE ISSUE OF ORGANICS IN DRINKING WATER

The general issue of the occurrence of persistent organic chemicals in the environment and their potential harmful effects on man has recently been raised regarding control of the level of organic compounds in finished drinking water. Recent concerns about organics in drinking water were expressed by Ongerth *et al.* (1973) in discussions concerning the suitability of wastewaters for consumptive reuse. These authors note that organic residues are found in wastewaters and water supplies but that little is known about their effects on man; they suggest that further research in this area is essential. Burnham *et al.* (1973) supported the position of Ongerth *et al.* (1973) and presented a new method for isolating and concentrating organic compounds in water employing polystyrene resins. Cameron *et al.* (1974) have emphasized that organic chemicals should be monitored in drinking supplies using the best available techniques. They also suggested that organochloride compounds were formed in water treatment plants as a consequence of normal treatment practices including prechlorination.

In early November 1974, the Environmental Defense Fund reported studies that suggested a possible link between some cancers and consumption of Mississippi River water by persons in Louisiana. EPA (1974) confirmed that a number of suspected carcinogens are present in the New Orleans water supply. Sixty-six compounds were identified including about 100 mg/l of chloroform, carbon tetrachloride and dieldrin—all suspected carcinogens. In November 1974, the results of an AWWA research committee on organic contaminants were published in the *Journal of the American Water Works Association.* The committee emphasized that little is known about the nature and harmful effects of organic chemicals in water and that analytical methods for such organics are inadequate. However, experimental studies with mice (Tardiff and Deinzer, 1973) showed that carbon chloroform extracts (CCE) and carbon alcohol extract (CAE) both contain toxicants, which was contrary to former beliefs that the CAE fraction was nontoxic.

The committee also defined a number of research needs based on panel discussions (Table IV).

In November 1974, the EPA announced that a nationwide survey would be conducted to determine the concentrations and potential harmful effects of organic chemicals in drinking water. In December of that same year, Bellar *et al.* (1974) reported on the occurrence of as much as 150 mg/l of chloroform in finished municipal water supplies and that these compounds are a result of the chlorination process. Repeated chlorine dosage was found to increase chloroform concentrations. These concentrations do not pose an acute hazard to man, but their chronic effect is unknown. During the same month the *Safe Drinking Water Act* was signed into law and included a provision to conduct comprehensive studies of the potential harmful impact of organic chemicals. Shortly thereafter, the National Organics Reconnaissance Survey (NORS) was announced, which listed 80 selected cities where studies of organic chemicals would be conducted.

Table IV. Research Needs Associated with Organics in Water

Identification of the types of organics present in natural and treated waters;

Origin and Fate of organics must be determined. The accumulation in sediments, uptake by aquatic biota, and nature of decomposition products in entire water basins must be understood;

Chelation abilities of organics should be studied;

Toxicity of organics, both acute and chronic, should be determined, as well as synergistic effects;

Epidemiological studies should be conducted to relate possible illnesses in the general population to various organics;

Monitoring techniques should be defined and should consider use of aquatic biota as indicators of contamination; and

Control of organics in water may be achieved by regulation of the discharge from sources, new treatment techniques at wastewater and water plants, or by modifications of chlorination practices at water and wastewater treatment plants. Research should be conducted to determine the most appropriate form of control.

The findings of another AWWA committee meeting on organics were published in the *Journal of the American Water Works Association* in August 1975. This committee determined that the concentrations of organics in raw water supplies has not decreased in recent years despite large-scale efforts to improve the efficiency of domestic and industrial waste treatment facilities. The committee recommended that the organic content of water supplies be monitored continuously, both before and after treatment, and that improvements in analytic techniques be investigated.

The results of the NORS for halogenated organics were reported in November 1975 by Symons *et al.* (1975). It was determined that four trihalomethanes (chloroform, bromodichloromethane, dibromochloromethane and bromoform) are widespread in finished drinking waters and result from chlorination. The concentration of these compounds in raw water was low.

The median and range of the observations in finished waters are chloroform (< 0.1, 21.0, 311.0 mg/l); bromodichloromethane (0, 6.0, 116.0 mg/l); dibromochloromethane (0, 1.2, 100.0 mg/l) and bromoform (0, 92.0 mg/l). Carbon tetrachloride was not widespread in the samples tested. It was determined that high concentrations of the halomethanes were associated with high levels of total organic carbon in the raw water when chlorine was added, and with plants that practice precipitative softening at high pH.

The Interim Primary Drinking Water Regulations were published in the *Journal of the American Water Works Association* in February 1976. These regulations call for a careful look at toxic chemicals including asbestos, PCBs and halogenated organics. This program includes requirements for organic chemical monitoring and an intensive research program dealing with the effects of organics, analytical procedures and treatment practices and technology.

Morris and Johnson (1976) have determined that high agricultural runoffs coincide with periods of high halomethane concentrations in finished drinking water in Iowa. It was found that turbidity removal prior to chlorination can help minimize chloroform production in plants. Finally, in November 1976, Stevens *et al.* (1976) used bench- and pilot-scale data to determine the influence of the concentration of precursor organics, pH, type of disinfectant and temperature on halomethane production. It was determined that the point of chlorination in the treatment process represents an important variable, which affects the ultimate halomethane concentrations in finished drinking waters.

EXAMPLE DDT MODEL

The AWWA Committee on organic chemicals and the EPA have raised questions regarding the fate and distribution of organics discharged into the environment from point and nonpoint sources. Resolution of these questions could help determine the relative importance of organics produced at water treatment plants during chlorination compared with other sources. This, in turn, would help determine the most efficient and cost-effective approach to the control of the total exposure of man to harmful organic chemicals, including those contained in finished drinking water.

The determination of the temporal and spatial distribution of organic contaminants in the environment at various trophic levels has been attempted using systems analysis and mathematical modeling techniques (Harrison *et al.*, 1970; Metcalf *et al.*, 1971; Woodwell *et al.*, 1971). Such models are based on the principle of mass continuity, which can be expressed by equations of the form:

$$V_j d \frac{(N_{ij}M_{ij})}{dt} = (K_{i-1,i}M_{i-1}N_{i-1})_jV_j - (K_{ii}M_iN_i)_jV_j$$

$$+ \text{ transport to adjacent segments} + \text{external loads} + \text{decay}$$

for a single homogeneous region in the environment (Hydroscience, 1973). N_{ij} is the mass of toxicant per mass of trophic level i in segment j; M_{ij} is the mass of trophic level i per unit volume of segment j; $K_{i-1,\ i}$ and K_{ii} represent the rate transfer of toxicant from the next lower and to the next higher trophic levels within segment j; and V_j is the volume of segment j. Solution of the above equation gives the mass of toxicant per unit biomass at each trophic level as a function of time and position within the environment. The first two terms account for bioaccumulation of toxicant in various trophic levels. The next term refers to mass transfers due to physical mechanisms such as advection, dispersion, evaporation, atmospheric fallout and flux across phase interfaces. Decay mechanisms include natural degradation caused by the metabolic activities of organisms. The external loads refer to inputs to the system from sources such as treatment plant effluents and landfill leachates.

The above equation can be applied to a variety of problems having different time and space scales. Time scales could range from daily or seasonal to long-term. Local spatial scales would normally be associated with daily fluctuations of contaminants. Regional spatial scales are often coupled with seasonal time scales, whereas long-term problems usually have crude and perhaps worldwide spatial extents.

As a crude example, Randers (1973) has developed a model for the global movement of DDT, which deals with the long-term distribution of DDT in the soil, air, rivers, oceans and marine fish (Figure 1). The purpose of the model is not to predict the future but rather to synthesize the best possible estimate of the probable impact of a control policy for DDT, given only bits and pieces of information currently available about the system. More specifically, the model was developed to determine how the costs and harmful impacts of a given pattern of DDT usage are distributed over time.

The model accounts for evaporation from the soil and biological degradation in the soil. Evaporation from the soil and precipitation from the air define mass transfers between these compartments. Both the air and soil are affected by inputs due to applications on soil and losses to the air. The mass of DDT in the rivers is increased by runoff from the soil and decreased by losses associated with particulate matter input to the oceans. The detention time within the world river systems is considered too small to permit significant decay. The ocean concentrations are affected by rainfall inputs; by death, uptake and excretions of fish; and by natural decay mechanisms. The quantity of DDT stored in fish flesh is affected by the amount in the ocean water and the concentration factor between ocean water and marine plankton.

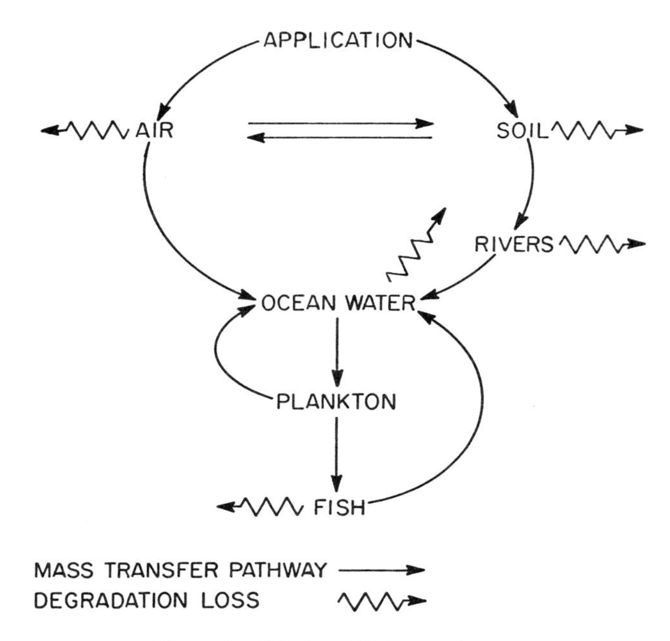

Figure 1. DDT model interactions.

The actual numerical value of the transfer and degradation rates in the model are still subject to some scientific uncertainty. Therefore, the effects of variation of the values of these coefficients associated with these processes was studied by Randers (1973) using the model. Table V lists the model coefficients and the best estimate of their numerical values.

Simulation results from the Randers model are shown in Figure 2. The model results are for a case where DDT use is gradually decreased up to the year 2000. The timing of the responses and the relative magnitude of the calculations are of more significance than the actual numbers. The results indicate that DDT levels in the soil and rivers follow closely the pattern of the application rate. However, because of numerous feedbacks and pathways in the model, DDT levels in the ocean and in fish continue to rise for several years despite control measures. Sensitivity studies by Randers (1973), with alternate coefficient values, show that the model trends are relatively unaffected by shifts in these numerical values, but that absolute values of the trends are highly dependent on some of the model coefficients. This suggests that the predictive value of the model is limited and dependent on additional field and laboratory studies, which may provide further insight into values of the model coefficients.

Table V. Estimated Coefficient Values for DDT Model

Model Parameter	Numerical Value
Airborne fraction (dimensionless)	0.5
Body weights eaten/yr (/yr)	10
Consumed fraction (dimensionless)	0.5
Degraded fraction (dimensionless)	0.1
Decay half-life in ocean (yr)	15
Decay half-life in soil (yr)	10
Evaporated half-life from soil (yr)	2
Excretion half-life from fish (yr)	0.3
Half-life of fish (yr)	3
Mass of fish (tons)	6×10^8
Mass of mixed layer (tons)	3×10^{16}
Ocean-plankton conc. factor	2000
Precipitation half-life (yr)	0.05
Runoff half-life (yr)	0.1
Soil fraction (dimensionless)	0.3
Solution half-life (yr)	500

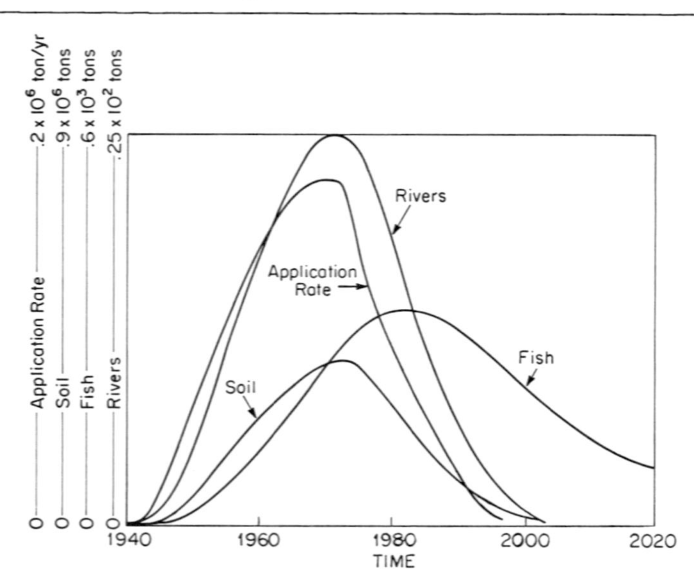

Figure 2. Simulation of the global distribution of DDT.

Despite acknowledged limitations of the model, certain policy implications are apparent. First, because of transport times and system delays, DDT concentrations in the ocean and fish respond several years following alterations of input rates to the land. The amounts of DDT in the rivers respond much more quickly to input changes. Second, DDT concentrations in marine food chains continue to increase several years after corrective action is taken to decrease the rate of application. This response pattern to changes in application rates has important engineering planning implications in terms of evaluating the total exposure of man to harmful organic chemicals such as DDT. For example, the results suggest that harmful organic chemicals, like DDT, in our water supplies will change relatively quickly following control measures, while organics in marine food supplies may continue at high levels despite corrective action. This, of course, suggests that in the long run, control of harmful organic chemicals at their source may be a more appropriate way to protect human health than relying solely on the effectiveness of water treatment plants. Furthermore, harmful chemicals in our food may become even more important relative to drinking water as more and more organic chemicals are regulated.

The above approach and model has only limited immediate applications to local or regional waste treatment and water supply issues. However, it is felt that such techniques can be successfully applied on a much smaller time and spatial scale as appropriate scientific and field monitoring data become available. It is felt that mathematical modeling techniques combined with the traditional scientific investigations offer the only hope for development of predictive tools, which can be used by policy-makers concerned with questions of toxic organic chemicals and the operation and placement of wastewater and water treament plants.

REFERENCES

Ahmed, A. K. "PCBs in the Environment," *Environment* 18(2):6-11 (1976).

AWWA Committee on Organic Contaminants in Water. "Organic Contaminants in Water," *J. Am. Water Works Assoc.* 66(11):682-688 (1974).

AWWA Committee on Organic Contaminants in Water. "Organic Contaminants in Water Supplies," *J. Am. Water Works Assoc.* 67(8):418-424 (1975).

Bellar, T. A., J. J. Lichtenberg and R. C. Kroner. "The Occurrence of Organohalides in Chlorinated Drinking Waters," *J. Am. Water Works.* 66(12):703-705 (1974).

Bunch, R. J., E. Barth and M. Ettinger. "Organic Materials in Secondary Effluents," *J. Water Poll. Control Fed.* 33(2):122 (1961).

Burnham, A. K., G. K. Calder, J. S. Fritz, G. A. Junk, H.J. Svec and R. Vick, "Trace Organics in Water: Their Isolation and Identification," *J. Am. Water Works Assoc.* 65(11):722-725 (1973).

Cameron, C. D., R. C. Hoehn and J. W. Strimaitis. "Organic Contaminants in Raw and Finished Water," *J. Am. Water Works Assoc.* 66(9):519-523 (1974).

Environmental Defense Fund. "The Implications of Caucer-Causing Substances in Mississippi River Waters," Washington, D.C. (1974).
Environmental Protection Agency. "New Orleans Water Supply Study," Slidell, Louisiana (1974).
Harrison, H. L., O. L. Loucks, J. W. Mitchell, D. F. Parkhurst, C. R. Tracy, D. G. Watts and V. J. Yannacone, Jr. "Systems Studies of DDT Transports," *Science* 170: 503-508 (1970).
Hayes, W. J., Jr. *Toxicology of Pesticides* (Baltimore, Md.: Williams & Wilkins, 1975).
Headley, J. C. and A. V. Kneese. "Economic Implications of Pesticide Use," *Ann N. Y. Acad. Sci.* 160:30-39 (1969).
Holden, A. V. "Monitoring Persistent Organic Pollutants," in *Organochloride Insecticides: Persistent Organic Pollutants*, F. Moriarty, Ed. (New York: Academic Press, 1975), pp. 1-27.
Hydroscience, Inc. "Limnological Systems Analysis of The Great Lakes," Westwood, N.J., DACW-35-71-C0030 (1973).
Metcalf, R. L., G. K. Sangha and I. P. Kapoor. "Model Ecosystem for the Evaluation of Pesticide Biodegradability and Ecological Magnification," *Envir. Sci. Technol.* 5(8):709-713 (1971).
Morris, R. L. and L. G. Johnson. "Agricultural Runoff as a Source of Halomethanes in Drinking Water," *J. Am. Water Works Assoc.* 68(9):492-494 (1976).
"National Interim Primary Drinking Water Regulations," *J. Am. Water Works Assoc.* 68(2):57-67 (1976).
Ongerth, J. J., D. P. Spath, J. Crook and A. E. Greenberg. "Public Health Aspects of Organics in Water," *J. Am. Water Works Assoc.* 65(7):495-498 (1973).
Painter, H. H., M. Viney and A. Bywaters. "Composition of Sewage and Sewage Effluents," *J. Inst. Sew. Purif.* 4:302 (1961).
Panel on Hazardous Trace Substances. "PCBs–Environmental Impact," *Environ. Res.* 5:249-362 (1972).
Randers. J. "DDT Movement in the Global Environment," in *Toward Global Equilibrium*, D. L. Meadows and D. H. Meadows, Eds. (Cambridge, Mass.: Wright Allen Press, 1973).
Rebhum, M. and J. Manka. "Classification of Organics in Secondary Effluents," *Environ. Sci. Technol.* 5(7):606 (1971).
Rosen, A. A., S. Katz, W. W. Pitt, Jr. and C. D. Scott. "The Determination of Stable Organic Compounds in Waste Effluents at Microgram per Liter Levels by Automatic High Resolution Ion Eschange Chromatography," *Water Res.* 6:1029 (1972).
Stevens, A. A., C. J. Slocum, D. R. Seeger and G. G. Robeck. "Chlorination of Organics in Drinking Water," *J. Am. Water Works Assoc.* 68(11): 615-620 (1976).
Symons, J. M., T. A. Bellar, J. K. Carswell, Jr., J. DeMarco, K. L. Kropp G. G. Robeck, D. R. Seeger, C. J. Slocum, B. L. Smith and A. A. Stevens. "National Organics Reconnaissance Survey for Halogenated Organics," *J. Am. Water Works Assoc.* 67(11):Part 1, 634-647 (1975).
Tardiff, R. G. and M. Deinzer. "Toxicity of Organic Compounds in Drinking Water," proceedings of the *15th Water Quality Conference*, Champaign, Illinois (1973).
Woodwell, G. M., P. P. Craig and H. A. Johnson. "DDT in the Biosphere: Where Does It Go?" *Science* 174:1101-1107 (1971).

5

REACTIONS OF TRACE ORGANICS WITH CHLORINE AND BROMINE

Frederic K. Pfaender

Assistant Professor of Environmental Microbiology
Department of Environmental Sciences and Engineering
School of Public Health
University of North Carolina
Chapel Hill, North Carolina

INTRODUCTION

For many years there has been concern about the organic compounds present in rivers, lakes, sewage effluents and drinking water. This interest centered on compounds used in agriculture and industry, especially pesticides and PCBs. Recently, this concern has shifted to halogenated organics produced as a result of chlorination processes used to control the microbial contamination of drinking water and wastewaters. It has been known for a long time that water chlorination results in the production of chlorinated organics, like the chlorophenols that can cause taste and odor problems in chlorinated waters (Donaldson, 1922; Enslow, 1934). Very recently, it was noted that that chloroform ($CHCl_3$) and other halogenated organics were present in the drinking water of New Orleans (Dowty *et al.*, 1975; EPA, 1974) and that consumption of this water was correlated with the incidence of cancer in male Caucasians (Page and Harris, 1974). At about this same time, Rook (1974) and Bellar *et al.* (1974) reported that chloroform and other chlorinated compounds were formed as a result of chlorination for disinfection. These findings led the Environmental Protection Agency to undertake a survey of organics in the drinking water of 80 cities throughout the United States (Symons *et al.*, 1975). The results of this study confirmed the ubiquitous distribution of trihalomethanes in drinking water and raised many questions about the precursors and reactions that lead to trihalomethane formation. More recent work by Rook (1976, 1977), and Stevens *et al.* (1976) has provided some of the answers, as will be discussed in this chapter, but a great deal of work has yet to be done to elucidate what organic constituents in water react with chlorine to produce chloroform.

Although the trihalomethanes are present in almost all drinking water, and have been shown to be carcinogenic, as will be discussed by Dr. Kraybill in Chapter 7, they are not the only type of chlorinated organic to be produced as a result of chlorination processes. There are excellent reports by several workers, of a vast array of chlorinated organic molecules that can be formed during aqueous chlorination (Carlson *et al.*, 1975; 1976; Glaze and Henderson, 1975; Glaze *et al.*, 1976; Jolley, 1973, 1975). Potentially, many of these chlorinated molecules may constitute an even greater problem than chloroform. These substances will also be discussed.

AQUEOUS CHLORINE CHEMISTRY

Although it is common to refer to aqueous chlorine as Cl_2 dissolved in water, a neutral pH chlorine is rapidly hydrolyzed according to the equation

$$Cl_2 + H_2O \rightarrow HOCl + H^+ + Cl^- \qquad (1)$$

with hydrolysis being complete in only a few seconds. Since hypochlorous acid behaves as a weak acid, it dissociates according to the equilibrium

$$HOCl \leftrightharpoons H^+ + OCl^- \qquad (2)$$

with the hypochlorite ion being predominant at pH above 7.5-7.8. Several other species of chlorine are also present including H_2OCl^+, Cl^+ and Cl^-_3, but usually in minute quantities.

In water containing ammonia there is a fairly rapid formation of chloramines:

$$NH_3 + HOCl \rightarrow NH_2Cl + H_2O \qquad (3)$$

$$NH_2Cl + HOCl \rightarrow NHCl_2 + H_2O \qquad (4)$$

$$NHCl_2 + HOCl \rightarrow NCl_3 + H_2O \qquad (5)$$

with NH_2Cl and $NHCl_2$ being more common at neutral pH values. The ability of chloramines to chlorinate organic compounds has not been extensively studied, but Morris (1967) feels that chloramine may be an effective chlorinating agent for phenols. In assessing the reactions of aqueous chlorine with organic materials, more than just the concentration of a given chlorine species must be considered. Of importance is the reactivity of the different chlorine species toward different organic substrates. The overall reactivity can be defined as the product of the concentration of a specific chlorine species and a reactivity term r, which is the reactivity of that species toward a specific substrate under one set of conditions:

$$\text{Reactivity} = R_i = r_i\ C_i \quad (6)$$

As chlorine concentrations and substrate levels change, so will the reactivity. In complex mixtures of organics, like sewage or most natural waters, the reactivities are summarized by

$$\text{Rate} = \Sigma R_i = \Sigma r_i\ C_i \quad (7)$$

Since r_i values are different for each different specific reactant, and change with conditions of pH, ion content and solution conditions, tables of relative specific reactivities cannot be compiled. Table I presents an example of an evaluation of relative reactivities compiled by Morris (1976)

Table I. Estimated Net Reactivities of Forms of Active Chlorine; pH 7, 15°C[a]

Species	Estimated Specific Reactivity	Fraction of Total Cl	Net Relative Reactivity
Cl_2	10^3	3×10^{-6}	0.003
HOCl	1	0.80	0.80
OCl^-	10^{-4}	0.20	0.00002
H_2OCl^+	10^5	10^{-8}	0.001

[a]From Morris (1976)

for several dilute chlorine species at pH 7. The valves are estimates based on reactivities of chlorine species toward nitrogenous compounds and are not meant to be used other than as an example. The order of the reactivities, however, will probably remain the same in any reactions that involve the electrophilic or oxidizing properties of the chlorine species, although the quantitive aspects may change. It therefore seems clear that at near neutral pH the major reactions of chlorine with organic compounds will involve HOCl as the chlorinating species. Under acidic solutions, reactions with Cl_2, H_2OCl^+, and even Cl^+ may be much more important, and in basic solutions dissociation to OCl^- is essentially complete. Since HOCl reacts with ammonia to produce chloramines, as well as with organics, it would be anticipated that lower quantities of chlorinated organics would be produced in waters with high ammonia concentrations.

Whenever Cl_2 or HOCl is added to water containing bromide ion there is a rapid formation of HOB_r according to the reaction

$$Br^- + HOCl \rightarrow HOBr + Cl^- \qquad (8)$$

The resulting HOBr is also an electrophilic agent and may react with organic compounds. Considerably less is known about the products formed in these reactions than in reactions with HOCl.

REACTION OF CHLORINE WITH ORGANIC COMPOUNDS

There are several types of reactions between organic compounds and HOCl in dilute aqueous solutions that should be considered. These include oxidation reactions, formation of N-chlorinated organics, addition of chlorine to olefinic bonds, substitution reactions and, finally, the haloform reaction, which is really a special case of a substitution-type reaction. In each case, one type reaction can be followed by additional reactions of the same or different type.

Oxidation Reactions

Oxidations may be the most common result of the reaction of HOCl with organics in aqueous solutions. Figure 1 presents examples of some

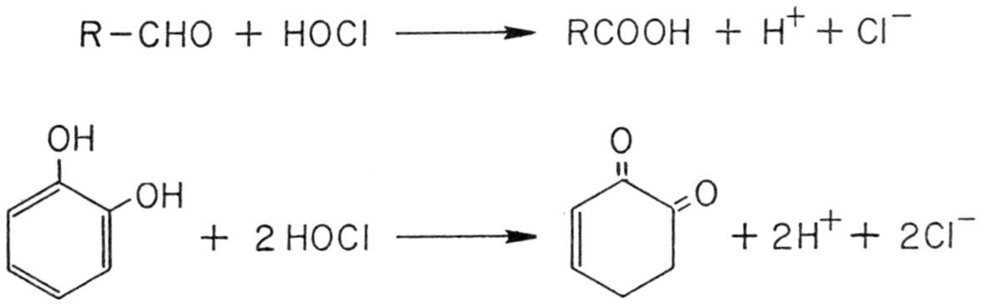

$$R{-}CHO + HOCl \longrightarrow RCOOH + H^+ + Cl^-$$

OH, OH (benzene ring) $+ 2\,HOCl \longrightarrow$ O, =O (ring) $+ 2H^+ + 2Cl^-$

Figure 1. Oxidation reactions involving HOCl and organic molecules.

of the reactions that can occur. These reactions do not occur with simple saturated aliphatic chains, but need a point of attack that may be a substitution or unsaturation in the molecule. Alcohols, aldehydes, carbohydrates, sulfhydryl groups or related sulfur linkages are all potential sites of oxidation. In natural waters and sewage streams, the carbohydrates and carbohydrate-related molecules are probably the most common substrate for oxidation reactions. Although these type reactions may account for a significant part of the chlorine demand exerted by natural waters and effluents, they do not involve the incorporation of chlorine into organic compounds; therefore, they will not be discussed further.

Formation of N-Chlorinated Organics

The chemistry of the formation of N-chlorinated compounds has been reviewed by Morris (1967). The reaction of HOCl with various nitrogenous compounds occurs according to the reaction

$$R_1 - \overset{H}{N} - R_2 + HOCl \rightarrow R_1 - \overset{Cl}{N} - R_2 + H_2O \qquad (9)$$

and can involve amines, amides, amino acids, proteins and heterocyclic compounds. The reactions proceed rapidly, especially with the more basic nitrogen atoms. With primary amines and ammonia, subsequent chlorinations can lead to di- and trichloramines and account for significant chlorine demand. Many nitrogeneous materials occur in natural and sewage effluents (Pitt *et al.,* 1975) that can potentially react to form N-chlorinated compounds. These reactions do not appear to result in the production of any C-chlorinated materials.

Addition of HOCl to Olefinic Bonds

The addition of hypochlorous acid to unsaturated bonds is illustrated by the type reaction

$$\gt C = C \lt + HOCl \rightarrow - \underset{Cl}{C} - \underset{OH}{C} - \qquad (10)$$

This reaction may occur in effluents containing unsaturated fatty acids and results in the production of chlorohydrins according to the reaction

$$CH_3\text{-}(CH_2)_6\text{-}C = C\text{-}(CH_2)_6\text{-}CH_3 + HOCl \rightarrow \begin{cases} CH_3\text{-}(CH_2)_6 - \underset{Cl}{\overset{H}{C}} - \underset{OH}{\overset{H}{C}} - (CH_2)_6 - CH_3 \\ CH_3\text{-}(CH_2)_6 - \underset{OH}{\overset{H}{C}} - \underset{Cl}{\overset{H}{C}} - (CH_2)_6\text{-}CH_3 \end{cases} \quad (11)$$

Reactions of this nature may also include the addition of chlorine to unsaturated sites on hydrocarbons and in the side chains of the phenylpropanoid units of lignin and possibly humic materials.

Substitution Reactions

The substitution of chlorine into organic compounds with the formation of C-chlorinated organics has been reviewed by Carlson *et al.* (1975), Jolley (1973) and Morris (1975). Two principal types of reactions have been observed—the substitution of chlorine into aromatic and heterocyclic ring systems and the haloform reaction, which produces trihalomethanes. The haloform reaction will be discussed in detail in a subsequent section. Figure 2 presents a type reaction for aromatic substitution and specific examples involving phenol and benzoic acid. Unsubstituted aromatic compounds react with chlorine only under strongly acidic conditions with H_2OCl^+ or Cl^+ being the actual chlorinating agent. However, the activation of the aromatic ring by the presence of an appropriate substituent such as hydroxyl or carboxyl groups allows the substitution of chlorine into the aromatic ring to proceed readily, even in neutral aqueous solutions (Morris, 1975). An example of chlorine uptake by various substituted aromatic substances is shown in Table II. Subsequent chlorination reactions can lead to di- and trichloroaromatics being produced, but Carlson and Caple (1976) report that in dilute aqueous solutions a limited number of chlorines are incorporated into aromatic rings. Natural waters and various municipal and industrial effluents contain myriad aromatic substances that can participate in substitution reactions with HOCl to produce chlorinated aromatic compounds. These compounds include phenols, aromatic carboxylic acids and the aquatic humic materials that contain many different aromatic moieties.

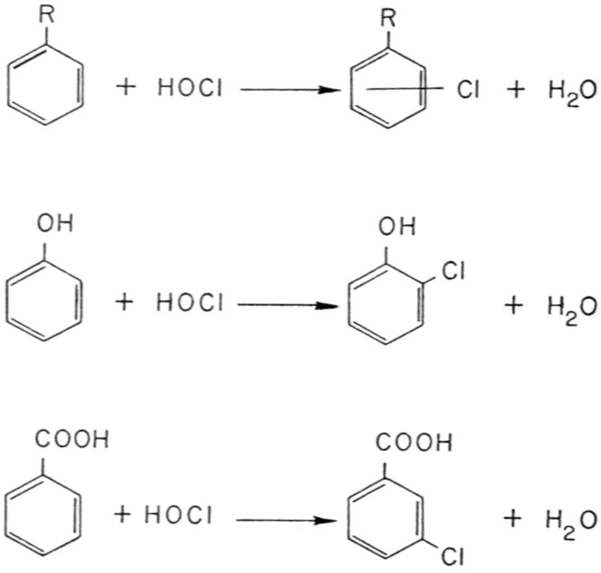

Figure 2. Aromatic substitution reactions with aqueous hypochlorous acid.

Table II. Uptake of Chlorine by Organic Compounds[a]

Compounds	% Chlorine Uptake		
	pH		
($9.5 \pm 0.6 \times 10^{-4}$ *M*)	3	7	10.1
Phenol	97.8	97.6	97.6
Anisole	80.7	11.4	2.8
Acetanilide	55.3	3.4	---
Toluene	11.1	2.9	---
Benzyl alcohol	2.3	---	---
Benzonitrile	2.1		
Nitrobenzene	1.8		
Chlorobenzene	1.8		
Methylbenzoate	1.8		
Benzene	1.5		

[a]From Carlson *et al.* (1975).

CHLORINATED ORGANIC COMPOUNDS PRODUCED IN WATER SYSTEMS AND SEWAGE

In the preceding section the various possible reactions between chlorine and organics were discussed. While there are many studies of the chemical reactions that might occur, there are far fewer investigations of the types of chlorinated compounds that do result from chlorination as used for treatment of water and wastewater. Most of the studies with drinking water have centered on chloroform, which appears to be the major chlorinated species produced, at least in terms of quantity. Much less attention has been directed toward the other chlorinated organics that may be formed during chlorination, although the NORS Study (Symons *et al.*, 1975) attempted to quantitate some of them. The very low concentrations (subpart per billion) of most of these compounds in treated water makes the analysis quite difficult.

The formation of chlorinated organics in sewage effluents disinfected by chlorination was studied in the classic work of Jolley (1973, 1975). Using radioactive ^{36}Cl, Jolley was able to show the incorporation of chlorine into at least 50 different constituents that were separable from one another by high-pressure liquid chromatography. Table III presents

Table III. Identification and Concentrations of Chlorine-Containing Constituents in Chlorinated Sewage Effluents[a]

Organic Compound	Concentration (μg/l)
5-chlorouracil	4.3
5-chlorouridine	1.7
8-chlorocaffeine	1.7
6-chloroquanine	0.9
8-chloroxanthine	1.5
2-chlorobenzoic acid	0.26
5-chlorosalicylic acid	0.24
4-chloromandelic acid	1.1
2-chlorophenol	1.7
4-chlorophenylacetic acid	0.38
4-chlorobenzoic acid	1.1
4-chlorophenol	0.69
3-chlorobenzoic acid	0.62
3-chlorophenol	0.51
4-chlororesorcinol	1.2

[a]From Jolley (1973).

a partial list of those compounds that were identified in the effluent, and their concentrations. Many of the chlorinated organics produced were not identified, but those that were include chlorinated purines, pyrimidines, phenols and aromatic acids. Of the initial chlorine applied, more than 99% was recovered as chloride ion, with about 1% being incorporated into organic compounds. This high chloride ion concentration would seem to indicate that oxidation reactions (as in Figure 1) constitute the major reaction mechanism occurring in sewage effluents, although some chloride may result from decomposition of chloramines. The other major reaction appears to be aromatic substitution reactions to produce the compounds identified.

Research on the production of chlorinated organics as a result of superchlorination of sewage effluents has been reported by Glaze and Henderson (1975) and Glaze *et al.* (1976). Using a microcoulometric procedure to detect total organically bound halogen, it was shown that organohalogen concentrations could increase by as much as one and one half orders of magnitude when high chlorine dose (2000 mg/ℓ) was applied to municipal wastewater (Glaze *et al.*, 1976). More than 50 compounds have been identified by gas chromatography/mass spectrometry, some of which are shown in Table IV. Both aromatic and nonaromatic compounds were produced, some containing multiple halogens. Most of these materials are present in the part per billion concentration range.

Table IV. Compounds Identified in Superchlorinated Municipal Effluents[a]

Nonaromatics	
Chloroform	Chlorocyclohexane
Dibromochloromethane	Tetrachloroacetone
Dichlorobutane	Pentachloroacetone
3-chloro-2-methylbut-1-ene	Hexachloroacetone
Aromatics	
o-Dichlorobenzene	Trichlorocumene
p-Dichlorobenzene	Dichlorotoluene
Chloroethylbenzene	Chlorocumene
Dichloroethylbenzene	Trichlorophenol
N-Methyl-trichloroaniline	Tetrachlorophenol
Trichloromethoxybenzene	Tetrachlorodimethoxybenzene
Tetrachloroethylstyrene	Trichlorophthalate
Trichloromethylstyrene	Tetrachlorophthalate
Trichloroethylbenzene	

[a]From Glaze and Henderson (1975).

Many organic compounds are present in natural waters (Christman and Ghassemi, 1966; Christman and Minear, 1971; Vallentyne, 1957), sewage effluents (Pitt *et al.,* 1974; Pitt *et al.,* 1975) and drinking water (EPA, 1975). Certainly some of these materials will be able to react with chlorine species to produce chlorinated organic molecules. Our concern at present is directed toward only a few of these molecules. They constitute only the tip of the iceburg, whose base we have yet to encounter.

HALOFORM REACTION

As already mentioned, the formation of chloroform in drinking water as a result of chlorination for disinfection was first reported by Rook (1974) and Bellar *et al.* (1974) at about the same time. It has been proposed that chloroform is produced by the classical haloform reaction, which has been known for many years. The reaction is between HOCl and an organic compound containing a methyl ketone group

$$CH_3\text{-}\overset{\overset{\displaystyle O}{\|}}{C}\text{-}$$

or some group, of CH_3-CHON- that can be oxidized to an acetyl group. The three hydrogens of the methyl group are sequentially replaced by chlorine according to the reaction mechanism (Morris, 1976).

$$R - \overset{\overset{\displaystyle O}{\|}}{C} - CH_3 \rightarrow R - \overset{\overset{\displaystyle O^-}{|}}{C} = CH_2 + H^+ \tag{12}$$

$$R - \overset{\overset{\displaystyle O^-}{|}}{C} = CH_2 + HOCl \rightarrow R - \overset{\overset{\displaystyle O}{\|}}{C} - CH_2Cl + OH^- \tag{13}$$

$$R - \overset{\overset{\displaystyle O}{\|}}{C} - CH_2Cl \rightarrow R - \overset{\overset{\displaystyle O^-}{|}}{C} = CHCl + H^+ \tag{14}$$

$$R - \overset{\overset{\displaystyle O^-}{|}}{C} = CHCl + HOCl \rightarrow R - \overset{\overset{\displaystyle O}{\|}}{C} - CHCL_2 + OH^- \tag{15}$$

$$R - \overset{\overset{\displaystyle O}{\|}}{C} - CHCl_2 \rightarrow R - \overset{\overset{\displaystyle O^-}{|}}{C} = CCl_2 + H^+ \tag{16}$$

$$R - \overset{\overset{\displaystyle O^-}{|}}{C} = CCl_2 + HOCl \rightarrow R - \overset{\overset{\displaystyle O}{\|}}{C} - CCl_3 + OH^- \tag{17}$$

The trihalomethane group undergoes nucleophilic base attack to release CCl^-_3, which combines with a hydrogen ion to produce chloroform.

$$R - \overset{\overset{\displaystyle C}{\|}}{C} \text{-} CCl_3 + OH^- \rightarrow R - \overset{\overset{\displaystyle O}{\|}}{C} - OH + HCCl_3 \tag{18}$$

The slowest step in the reaction is the initial enolization (Equation 12) and, therefore, the ultimate rate of chloroform formation is dependent on how fast the first reaction occurs. The reaction rate is not influenced by which halogenating agent participates in Equations 13, 15 and 17; the reaction rate is determined by Equation 12. The formation of the enol is favored by high pH, as substantiated by the several observations of increased chloroform production at increased pH (Rook, 1976; Stevens *et al.,* 1976). In addition to reactions with HOCl, it is known that HOBr can participate as the halogenating agent, and in fact, $CHCl_2Br$, $CHClBr_2$ and $CHBr_3$ are commonly observed in water systems (Symons *et al.,* 1975) and the occurrence of three iodomethanes, $CHCl_2I$, CHClBrI and $CHBr_2I$ in drinking water has been reported (Glaze *et al.,* 1976).

CHLOROFORM PRECURSORS

Stevens *et al.* (1976) have studied the chlorination of acetone as a representative of simple methyl ketones. Acetone does produce chloroform under water treatment-type parameters, but not in sufficient quantities to be able to account for the chloroform concentrations observed in treated waters. The reaction with acetone is both pH- and concentration-dependent, producing low concentration of chloroform at neutral pH values and the nonvolatile total organic carbon (NVTOC) levels of raw drinking water supplies. To account for the chloroform concentrations observed in drinking water being produced from acetone, the pH would have to be 10.2, which is rare, except perhaps in lime-softening operations; or the NVTOC levels would have to be over 10 mg/ℓ, which is also rare. This data would tend to indicate that in normal drinking water treatment practices, acetone and most methyl ketones (Morris, 1975) are probably not major contributors in the formation of chloroform. This however, does not preclude the possibility that under the high pH conditions of some treatment operations or in some industrial wastes with high carbon content, that methyl ketones might not produce significant amounts of chloroform.

If methyl ketones are not the source of chloroform, then what is? The data from the NORS study (Symons *et al.,* 1975) indicates that trihalomethanes are ubiquitous pollutants and, therefore, must either be produced from many different materials, or the precursors must also be ubiquitously distributed. At present, a clear choice between these possibilities cannot be made, and indeed may not need to be, for both could occur. The search for a widely distributed organic precursor has led to the examination of humic acid. Stevens *et al.* (1976) showed that humic acid solutions produced chloroform when subjected to drinking water level chlorination procedures. Figure 3 clearly demonstrates that at carbon concentrations close to those found in raw drinking water and

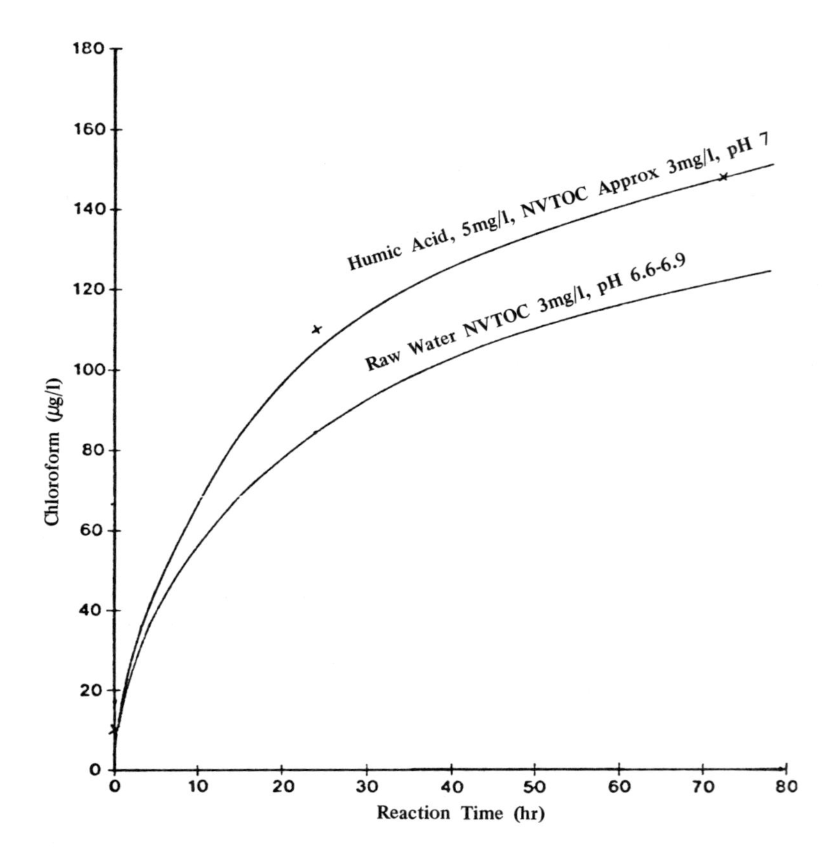

Figure 3. Comparison of humic acid, raw water reaction rates at similar NVTOC concentrations. Chlorine dose is 10mg/ℓ (Stevens *et al.*, 1976).

under actual practice conditions of pH and chlorine dose, humic acid solutions produce chloroform concentrations close to those observed in raw water. The reaction of humic acid to produce chloroform during chlorination responds to temperature and pH changes in a manner identical to changes observed with raw-water chlorination. Rook (1974) chlorinated peat extracts and found chloroform produced. In a subsequent paper (Rook, 1976), some specific molecular structures were examined for their ability to produce chloroform. Table V presents Rook's data, which show that *m*-dihydroxy aromatics (resorcinol) and 1,3-diketocyclohexanes (indandione, dimedon) produce chloroform in high yields

Table V. Chloroform Yields (%) from 4-hour Chlorination at 10°C[a]

Compound	pH 7.5	pH 11
Resorcinol	75	80
Catechol	trace	5
Hydroquinone	trace	26
Pyrogallol	trace	5
Phloroglucinol	---	1.5
1,3-indandione	58	100
Dimedone	85	100
1,3-Cyclohexanedione	50	71
1,3-Cyclohexanediol	0	0
1,2-Cyclohexanedione	0	0
1,4-Cyclohexanedione	0	14
Phenol	---	1
Acetone	trace	6.6
Ethanol	0	0

[a]From Rook (1976).

while orthodihydroxy aromatics and other type substitution patterns on cyclohexane do not. Also demonstrated is the expected stimulation of the reaction by basic conditions. The data of Rook and Stevens both point to humic material in water as a major precursor of chloroform.

Humic acid is a functional term used to describe a complex mixture of organic compounds. Soil humic material is thought to be formed from the microbial degradation products of lignin and other wood-derived materials, as well as polypeptides and carbohydrates, which combine in an as yet unknown manner (Schnitzer and Khan, 1972). The humic material found in water is different in some characteristics from soil humics, including higher fulvic acid content and more carboxyl acidity, but appears to retain the same polyaromatic composition that characterized all humics. Most of what is known about the specific molecules present in aquatic humics is based on the work of Christman and Ghassemi (1966), who isolated and identified the first seven structures shown in Figure 4 from a CuO degradation of concentrated aquatic humics. These seven molecules may represent only a small fraction of the diverse compounds that may be present. As shown by the data of Rook and from recent experiments in our own laboratory (Christman, Johnson and Pfaender, unpublished data), resorcinol is readily chlorinated, producing monochloro-, dichloro- and trichlororesorcinol, as well as chloroform. The degradation products of aquatic humics include resorcinol and other *m*-dihydroxy aromatics that may also produce chloroform. In addition, there may be reactions involving side chains, like those of the phenylpropane units also shown in Figure 4. These two phenylpropanes

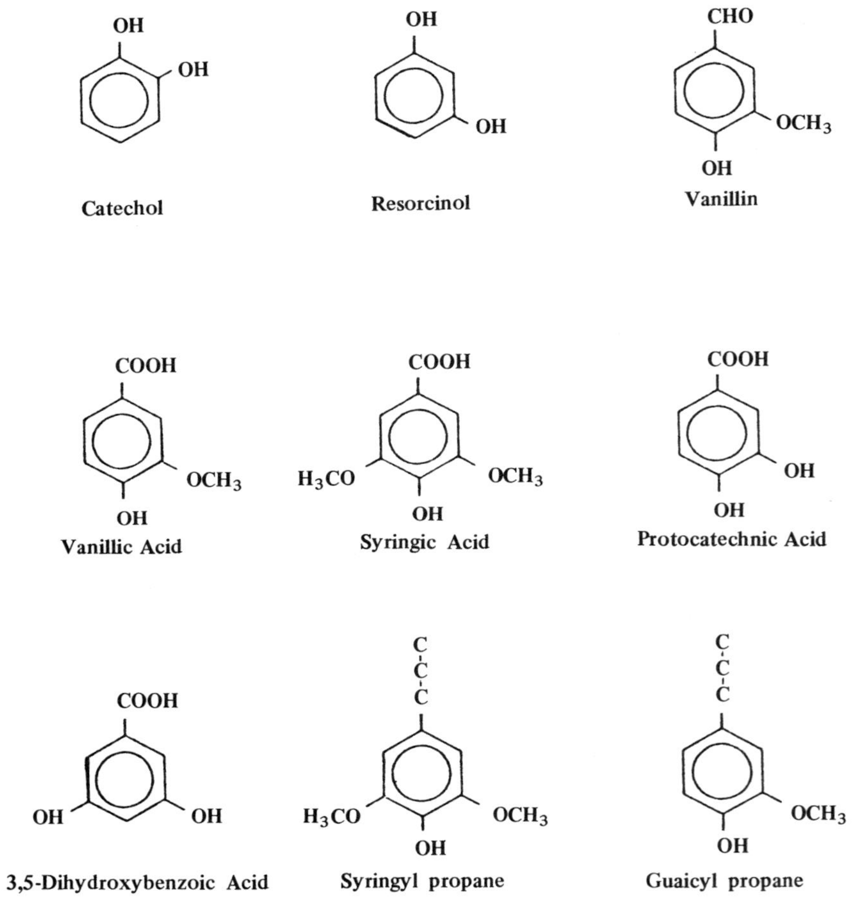

Catechol Resorcinol Vanillin

Vanillic Acid Syringic Acid Protocatechnic Acid

3,5-Dihydroxybenzoic Acid Syringyl propane Guaicyl propane

Figure 4. Degradation products of aquatic humics and lignin monomers.

represent the major structural subunits of lignin (Sarkanen and Ludwig, 1971), whose presence in humic material has not yet been confirmed. The three carbon side chains often contain oxidized and unsaturated sites that could participate in a haloform-type reaction. At present, resorcinol is the only known component of aquatic humics that has been shown to produce chloroform, but there are many other known or suspected structures that could theoretically react with HOCl to produce chloroform or other chlorinated organics. A major research need is more specific information about the reactions of humic material components and HOCl during water chlorination. Also needed is more knowledge of the specific molecules that are part of the humic polymer and are potentially available to react with chlorine. While there seems little doubt that humic material is a major contributor to the chloroform produced during water chlorination, the specific precursors involved are, for the most part, still unknown.

SUMMARY AND CONCLUSIONS

The reaction of organic matter and chlorine used for disinfection has been shown to produce compounds like chloroform, that may constitute a health hazard. The major reactive species of chlorine in dilute aqueous solution at neutral pH is HOCl. Chlorine can react with organic materials in a number of ways including (a) oxidation, (b) reactions with nitrogen, (c) addition to olefinic bonds, and (d) substitution reactions. The latter three reactions involve the incorporation of chlorine into organic compounds and the last two, the formation of carbon-chloride compounds. The examination of drinking water and wastewaters has shown that the production of many chloro-organic compounds can result from chlorination and that chloroform, while usually present in highest concentration, does not represent the sole compound about which we must be concerned. The haloform reaction involves the sequential incorporation of chlorine into the enol configuration of a methyl ketone. The overall haloform reaction rate is determined by the rate at which the enol is formed and is independent of the nature of the halogenating agent. The specific organic constituent(s) in natural water that reacts with chlorine to produce chloroform is not known for certain, but experimental evidence indicates that humic material may be the major contributor. Simple methyl ketones are unlikely to be significant precursors for the production of chloroform at the concentration and pH existing in natural waters. Many organic compounds are known to be present in natural waters, drinking water and wastewaters that can react with aqueous chlorine to produce a variety of chloro-organics, one of which is chloroform. To a large extent, the nature of the reactants, mechanisms of the reactions and potential health effects of the chlorinated products are about equally unknown.

REFERENCES

Bellar, T. A., J. J. Lichtenberg and R. C. Kroner. "The Occurrence of Organohalides in Finished Drinking Water," *J. Am. Water Works. Assoc.* 66:703-706 (1974).

Carlson, R. M. and R. Caple. "Organochemical Implications of Water Chlorination," in *The Environmental Impact of Water Chlorination*, R. L. Jolley, Ed. (Oak Ridge, Tenn.: Oak Ridge National Laboratory, 1976).

Carlson, R. M., R. E. Carlson, H. L. Kopperman and R. Caple. "Facile Incorporation of Chlorine into Aromatic Systems During Aqueous Chlorination Processes," *Environ. Sci. Technol.* 9:674-675 (1975).

Christman, R. F. and M. Ghassemi. "Chemical Nature of Organic Color in Water," *J. Am. Water Works Assoc.* 58:723-741 (June 1966).

Christman, R. F. and R. A. Minear. "Organics in Lakes," in *Organic Compounds in Aquatic Environments,* S. D. Faust and J. V. Hunter, Eds. (New York: Marcel Dekker, Inc., 1971).

Donaldson, W. "Observation on Chlorination Tastes and Odors," *Eng. Cont., Waterworks Monthly Issue,* 74-78 (1922).

Dowty, B., D. Carlisle, J. C. Laseton and J. Storer. "Halogenated Hydrocarbons in New Orleans Drinking Water and Blood Plasma," *Science* 187:75-77 (1975).

Enslow, L. H. "Modern Water Chlorination Practice," *J. Northeast Waterworks Assoc.* 48:6-22 (1934).

Glaze, W. H. and J. E. Henderson IV. "Formation of Organochlorine Compounds from the Chlorination of a Municipal Secondary Effluent," *J. Water Poll. Control Fed.* 47:2511-2515 (1975).

Glaze, W. H., J. E. Henderson IV and G. Smith. "Analysis of New Chlorinated Organic Compounds Formed by Chlorination of Municipal Wastewater," in *The Environmental Impact of Water Chlorination,* R. L. Jolley, Ed. (Oak Ridge, Tenn.: Oak Ridge National Laboratory, 1976).

Jolley, R. L. "Chlorination Effects on Organic Constituents in Effluents from Domestic Sanitary Sewage Treatment Plants," Ph.D. Dissertation, University of Tennessee (1973).

Jolley, R. L. "Chlorine Containing Organic Constituents in Sewage Effluents," *J. Water Poll. Control Fed.* 47:601-608 (1975).

Morris, J. C. "Kinetics of Reactions Between Aqueous Chlorine and Nitrogen Compounds," in *Principles and Applications of Water Chemistry,* S. D. Faust and J. V. Hunter, Eds. (New York: John Wiley and Sons, Inc., 1967).

Morris, J. C. "Formation of Halogenated Organics by Chlorination of Water Supplies," EPA-600/1-75-002 (Washington, D.C.: U.S. Environmental Protection Agency, 1975).

Morris, J. C. "The Chemistry of Aqueous Chlorine in Relation to Water Chlorination," in *The Environmental Impact of Water Chlorination,* R. L. Jolley, Ed. (Oak Ridge, Tenn.: Oak Ridge National Laboratory, 1976).

Page, T. and R. H. Harris. "Implications of Cancer-Causing Substances in Mississippi River Water," Environmental Defense Fund Report, Washington, D.C. (1974).

Pitt, W. W., R. L. Jolley and S. Katz. "Automated Analysis of Individual Refractory Organics in Polluted Water," EPA-660/2-74-076 (Washington, D.C.: U.S. Environmental Protection Agency, 1974).

Pitt, W. W., R. L., Jolley and C. D. Scott. "Determination of Trace Organics in Municipal Sewage Effluents and Natural Waters by High Resolution Ion-Exchange Chromatography," *Environ. Sci. Technol.* 9:1068-1073 (1975).

Rook, J. J. "Formation of Haloforms During Chlorination of Natural Waters," *J. Water Treat. and Exam.* 23:234-243 (1974).
Rook, J. J. "Haloforms in Drinking Water," *J. Am. Water Works Assoc.* 68:168-172 (1976).
Rook, J. J. "Chlorination Reactions of Fulvic Acids in Natural Waters," *Environ. Sci. Technol.* (In Press).
Sarkanen, K. V. and C. H. Ludwig, Eds. *Lignins* (New York: Wiley-Interscience, 1971).
Schnitzer, M. and S. U. Khan. *Humic Substances in the Environment* (New York: Marcel Dekker, Inc., 1972).
Stevens, A. A., C. J. Slocum, D. R. Seeger and G. G. Robeck. "Chlorination of Organics in Drinking Water," in *The Environmental Impact of Water Chlorination,* R. L. Jolley, Ed. (Oak Ridge, Tenn.: Oak Ridge National Laboratory, 1976).
Symons, J. M., T. A. Bellar, J. K. Carswell, J. DeMarco, K. L. Kropp, G. G. Robeck, D. R. Seeger, C. J. Slocum, B. L. Smith and A. A. Stevens. "National Organics Reconnaissance Survey for Halogenated Organics in Drinking Water," *J. Am. Water Works Assoc.* 67:634-647 (1975).
United States Environmental Protection Agency. "New Orleans Water Supply Study, Lower Mississippi River Facility," draft analytical report, Slidell, Louisiana (1974).
United States Environmental Protection Agency. "Suspect Carcinogens in Water Supplies," Office of Research and Development, Interim Report with Appendixes, Washington, D.C. (1975).
Vallentyne, J. R. "The Molecular Nature of Organic Matter in Lakes and Oceans with Lesser Reference to Sewage and Terrestrial Soils," *J. Fish. Res. Bd. Can.* 14:33-82 (1957).

6

REACTIONS OF OZONE WITH TRACE ORGANICS IN WATER AND WASTEWATER

L. Joseph Bollyky

Bollyky Associates
Stamford, Connecticut

Ozone has been used effectively since the turn of the century for the removal of taste, color and odor from drinking water (Bollyky, 1971, 1974, 1976, 1977). There are well over 1000 municipal water treatment plants that utilize the high oxidizing power of ozone for the removal of trace organics from water throughout the world. Over 30 of these plants are located in French-speaking Canada, and there are four in the U.S. including the two plants currently under construction in Michigan at Monroe and Bay City.

In the industrial wastewater treatment field ozone is used effectively and economically for the removal of low concentrations of organic compounds such as cyanides and phenols which are not readily treatable by biological oxidation (Bollyky *et al.*, 1976; Gould and Weber, 1976). Nonbiodegradable organics can be made biodegradable by ozone treatment alone or in the presence of UV light (Prengle *et al.*, 1977).

There is a set of convincing experimental evidence available indicating that the humic acid precursors of chlorinated organics can be removed from water by a combination of ozone treatment and activated carbon biological filter (Kuhn *et al.*, 1977). Here the role of ozone is to convert the poorly biodegradable organics into more readily biodegradable partially oxidized products which undergo biological degradation on the carbon filter. This biological process is strongly assisted by the high dissolved oxygen content of the water after the ozone treatment.

The ozone- and activated carbon-treated water shows a surprisingly low bacteria count (Kuhn *et al.*, 1977). The replacement of carbon is much less frequent than would be expected for purely adsorption-type carbon treatment.

Ozone alone is capable of oxidizing practically any organic compound to carbon dioxide (Heist, 1973). However, the reaction may be slow and may require large amounts of ozone. When ozone is used in combination with UV light, the reaction is rapid but still requires large amounts of ozone (Gould and Weber, 1976). In the ozone-activated carbon process a relatively small amount of ozone is used since most of the oxidation occurs during the biological process.

OZONE

It is a gas, a triatomic allotrope of oxygen produced from air or oxygen by an ozone generator at the site of application. The output of the ozone generator typically contains up to 1-4% ozone in the unreacted air or oxygen.

Ozone is a very powerful oxidizing agent and a very powerful disinfectant which does not leave undesirable residues behind. The oxidizing power of a chemical usually parallels its disinfecting power, and they both are reflected in its oxidation potential. As Table I indicates, the oxidation potential of ozone is E = 2.08 volts, the highest among the commonly available chemicals suitable for water treatment.

Table I. Oxidation Potentials of Chemical Disinfectants

		Oxidation Potentials (E°_{ox} volts)
Ozone	$O_3 + 2H^+ + 2e^- \rightarrow O_2 + H_2O$	2.07
Permanganate	$MnO^- + 4H^+ + 3e^- \rightarrow MnO_2 + 2H_2O$	1.67
Hypobromous Acid	$HOBr + H^+ + e^- \rightarrow 1/2\ Br_2 + H_2O$	1.59
Chlorine Dioxide	$ClO_2 + e^- \rightarrow ClO^-$	1.50
Hypochlorous Acid	$HOCl + H^+ + 2e^- \rightarrow Cl^- + H_2O$	1.49
Hypoiodous Acid	$HOl + H^+ + e^- \rightarrow 1/2\ I_2 + H_2O$	1.45
Chlorine Gas	$Cl_2 + 2e^- \rightarrow 2Cl^-$	1.36
Bromine	$Br_2 + 2e^- \rightarrow 2Br^-$	1.09
Iodine	$I_2 + 2e^- \rightarrow 2I^-$	0.54

Since ozone is an unstable and very reactive compound, it is either consumed by the oxidation of dissolved organics and other oxidizable materials or decomposes thermally reverting back to oxygen. In any event it does not produce a long-lasting residual. The oxidation products are usually nontoxic and harmless to either humans or to marine life.

The cost of ozone is relatively high (20¢ to 35¢/lb) depending on the cost of electricity, the capacity of the ozone generator used, and whether air or oxygen is used as feed gas. This high cost is primarily responsible for the fact that ozone has not been used very widely for water treatment in the U.S. However, with improvements in ozone generator technology, the cost is expected to decrease. The lower cost, in turn, should broaden the use of ozone for water and wastewater treatment.

OZONE CONTACTORS

The mass transfer of ozone from the ozone-air or ozone-oxygen mixture into the water requires specially designed gas-liquid contactors (Bollyky and Nebel, 1977; Masschelain *et al.,* 1975; Stahl, 1975). General requirements are maximum transfer of ozone into the water with minimum head loss of the water and maximum flexibility for variation in the flow of the water.

The most frequently used method is the diffusion of the ozone-air mixture into a minimum of 16 feet of water (see Figure 1). The ozone-air mixture is pressed through 20-60 permeability porous ceramic diffusers to generate fine bubbles at the bottom of the contact tank. The fine bubbles will then rise usually countercurrent to the water flow, and a 90-98% portion of the ozone is transferred into the solution. The off-gases are collected and usually reintroduced into the incoming raw water to transfer the remaining portion of the ozone into the water. The off-gases from this second pass still contain a little ozone. However, these remaining few parts per million can be removed by passing the gas through an ozone decomposer.

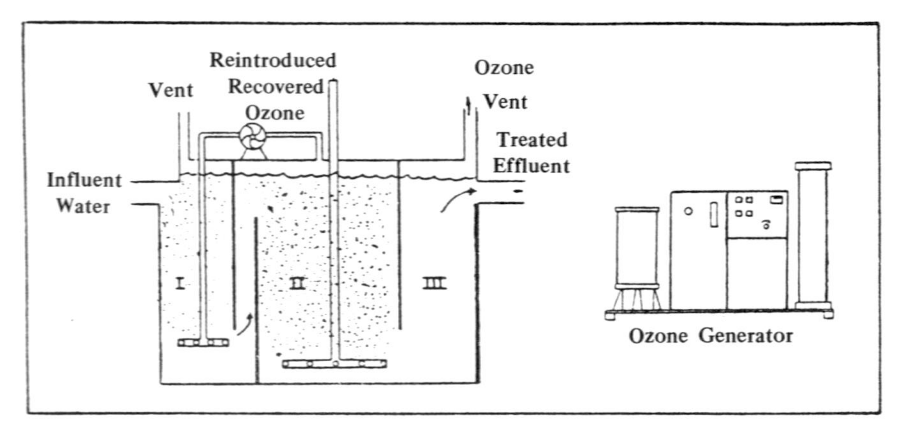

Figure 1. Ozone contact chamber for large flows of water. Chambers: I. Removal of ozone from off-gas, contact time 2 to 3 min. II. Main treatment, water depth 14 ft, contact time 4 to 6 min. III. Contact time 2 to 3 min for water and 10 min for wastewater. Water depth: 14 to 18 ft.

The large-scale diffusion-type ozone contactors are usually inground, baffled and covered concrete tanks (Figure 1). The center portion (Chamber II) is where ozone is first introduced into the water. Chamber I is used for the reintroduction of the off-gases into the incoming water. Finally, the third section (Chamber III) is used for the removal of excess air and ozone bubbles from the water. The overall ozone transfer might be 99+% efficient.

REACTIONS

The reactions of ozone with organic compounds fall into four major groups as follows:

1. The reaction with organic compounds containing double bonds (olefins) involves the formation of a pi complex between ozone and the olefin which either proceeds into 1, 3-dipolar cyclo addition and ozonalysis products or forms a sigma complex which loses oxygen to produce an epoxide (Am. Chem. Soc., 1972; Oehlschlaeger, 1977). Diperoxides and polymeric peroxide may also form (see Figure 2). A commercially important reaction n of ozone of this type is the conversion of oleic acid into pelargonic acid and azealic acid as shown in Figure 3. Some of the other olefin reactions are shown in Figure 4.

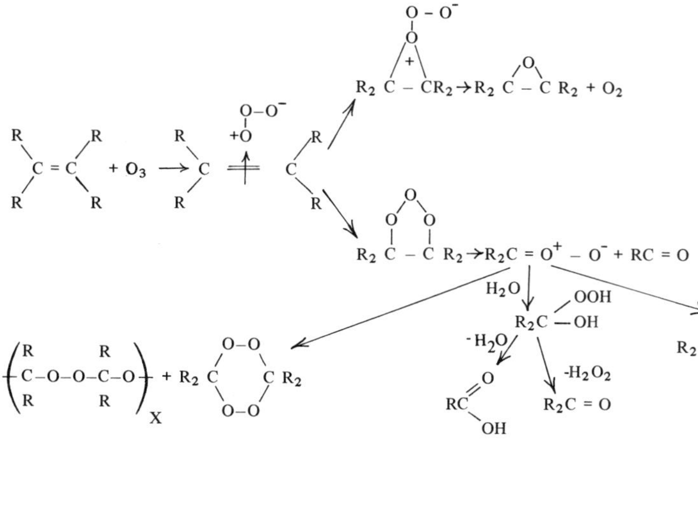

Figure 2. Mechanism of ozone-olefin reactions.

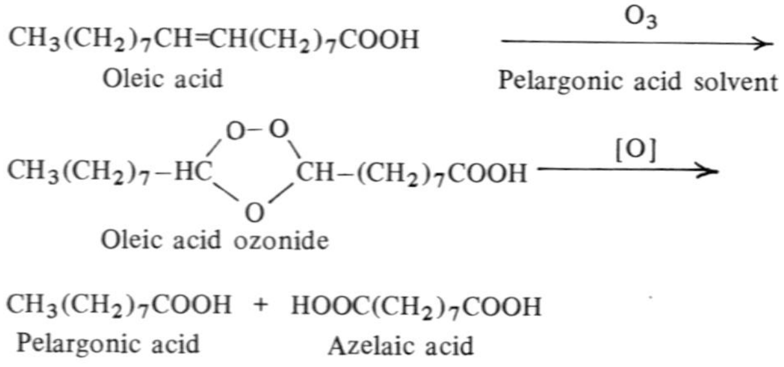

$$CH_3(CH_2)_7CH{=}CH(CH_2)_7COOH \xrightarrow[\text{Pelargonic acid solvent}]{O_3}$$

Oleic acid

$$CH_3(CH_2)_7{-}HC\overset{O-O}{\underset{O}{}}CH{-}(CH_2)_7COOH \xrightarrow{[O]}$$

Oleic acid ozonide

$$CH_3(CH_2)_7COOH + HOOC(CH_2)_7COOH$$

Pelargonic acid   Azelaic acid

Figure 3. Ozonolysis of oleic acid.

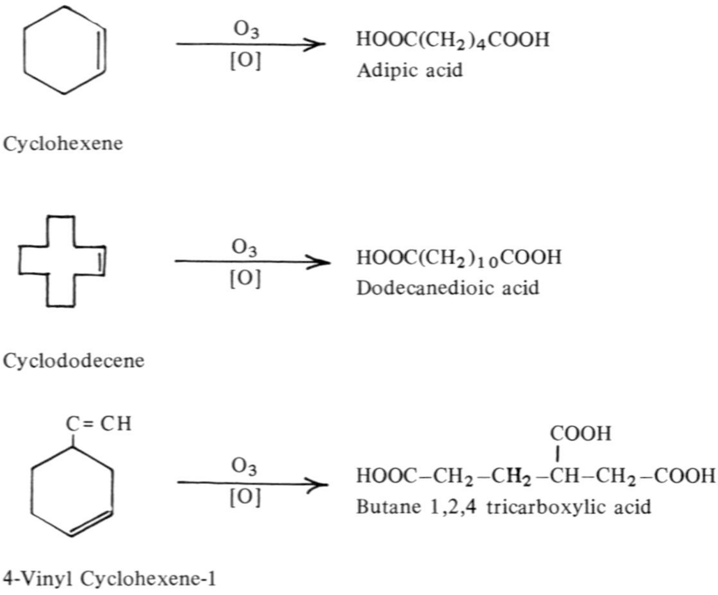

Cyclohexene $\xrightarrow[{[O]}]{O_3}$ $HOOC(CH_2)_4COOH$ Adipic acid

Cyclododecene $\xrightarrow[{[O]}]{O_3}$ $HOOC(CH_2)_{10}COOH$ Dodecanedioic acid

4-Vinyl Cyclohexene-1 (C = CH) $\xrightarrow[{[O]}]{O_3}$ $HOOC{-}CH_2{-}CH_2{-}\underset{}{CH}(COOH){-}CH_2{-}COOH$ Butane 1,2,4 tricarboxylic acid

Figure 4. Ozonolysis of cyclic olefins.

2. The reaction with nucleophiles is rapid and produces their oxides as shown in Figure 5. The compounds of this group frequently encountered in water treatment include cyanide, hydrogen sulfide, mercaptans, sulfoxide, amines and many nitrogen, sulfur and phosphorus compounds as indicated by Figures 6 and 7.

$$R_2S + O_3 \rightarrow R_2SO + O_2 \xrightarrow{O_3} R_2SO_2 + O_2$$

$$R_3N + O_3 \rightarrow R_3N \rightarrow O + O_2$$

$$R_3As + O_3 \rightarrow R_3As \rightarrow O + O_2$$

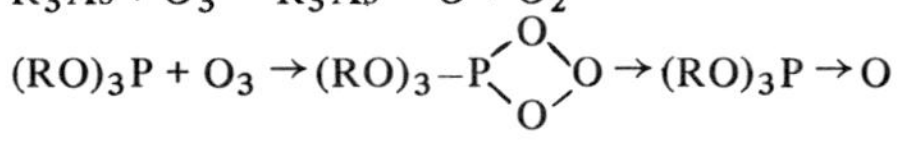

$$(RO)_3P + O_3 \rightarrow (RO)_3{-}P\langle O{-}O{-}O \rangle \rightarrow (RO)_3P \rightarrow O$$

Figure 5. Ozonation of nucleophiles.

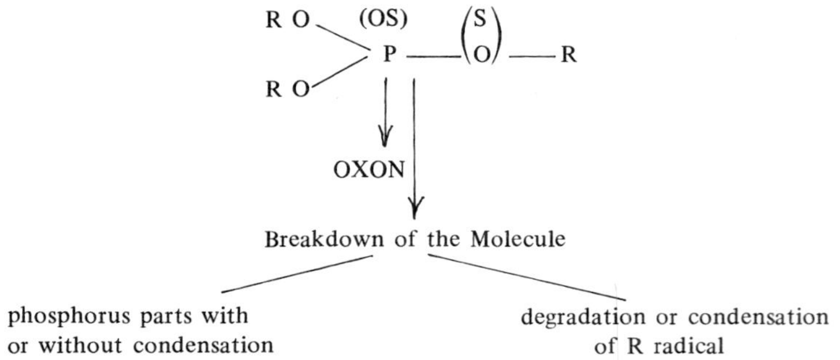

$(RO)_2P(OS){-}(S/O){-}R$

↓ OXON

↓ Breakdown of the Molecule

phosphorus parts with or without condensation

degradation or condensation of R radical

Figure 6. Ozone oxidation of organophosphorus insecticides.

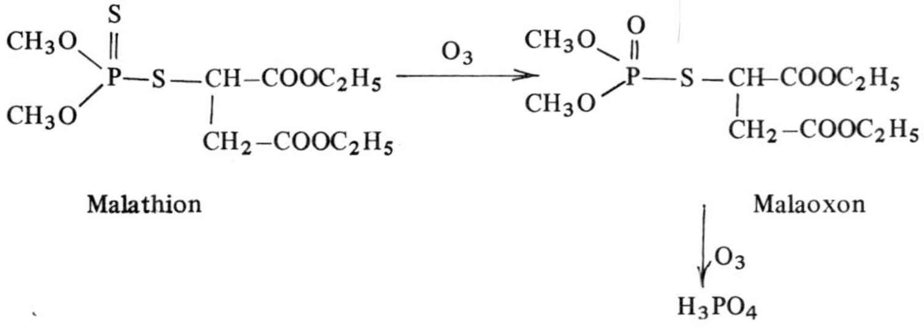

$$(CH_3O)_2P(=S){-}S{-}CH(COOC_2H_5){-}CH_2{-}COOC_2H_5 \xrightarrow{O_3} (CH_3O)_2P(=O){-}S{-}CH(COOC_2H_5){-}CH_2{-}COOC_2H_5 \xrightarrow{O_3} H_3PO_4$$

Malathion → Malaoxon → H_3PO_4

Figure 7. Ozonation of malathion.

The reaction of malathion is an unusual case. Although the products of ozone oxidation reactions are most frequently less toxic than the starting material, the first intermediate oxidation product of malathion is malaoxon, which is a more toxic compound than malathion. It reacts further, however, to produce phosphoric acid, a harmless product (Richard and Brener, 1977).

3. The reaction with aromatic compounds, such as benzene, fragments the ring structure as shown in Figures 8 and 9. A much-studied reaction is the oxidation of phenol (Prengle *et al.*, 1977). The reactions proceed through several intermediates as shown in Figure 10. The disappearance of phenol and the formation of intermediates is shown as a function of reaction time in Figure 11. The total organic carbon removal is shown in Figure 12 as the function of ozone dosage. Note the amount of ozone necessary for complete oxidation to carbon dioxide.

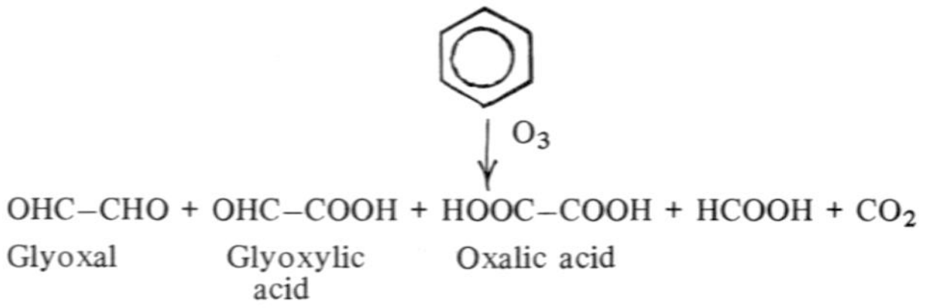

O_3

$OHC{-}CHO + OHC{-}COOH + HOOC{-}COOH + HCOOH + CO_2$

Glyoxal — Glyoxylic acid — Oxalic acid

Figure 8. Ozonolysis of benzene.

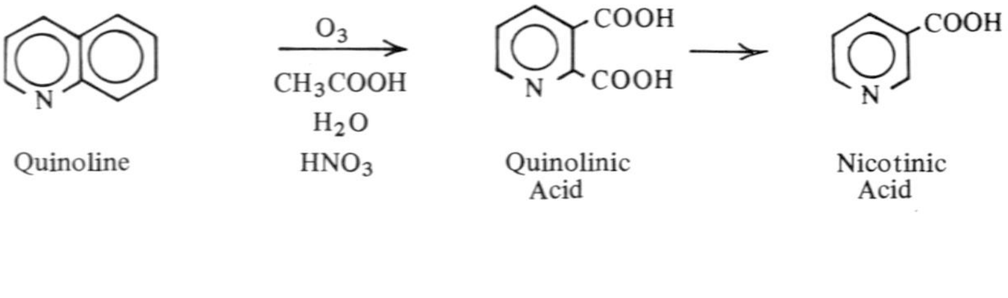

Quinoline $\xrightarrow[CH_3COOH,\ H_2O,\ HNO_3]{O_3}$ Quinolinic Acid (N, COOH, COOH) $\longrightarrow$ Nicotinic Acid (N, COOH)

Figure 9. Ozonolysis of quinoline.

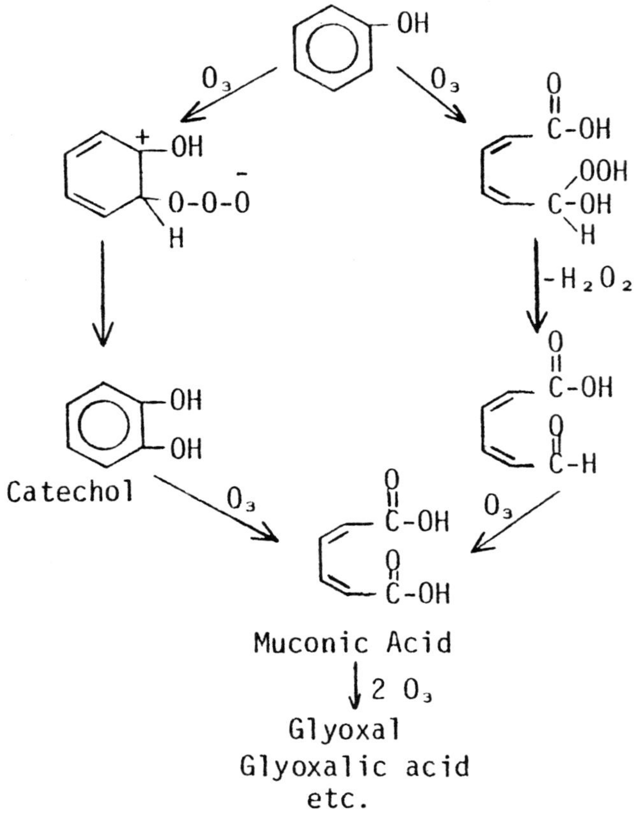

Figure 10. Ozonolysis of phenol in water.

4. The reaction with aliphatic alcohols, acids and hydrocarbons is relatively slow, but it becomes substantially faster in the presence of UV light, as indicated in Figures 12, 13, 14, 15 and 16 (Kuo *et al.*, 1977). The combination of ozone and UV is a very powerful oxidizing environment. Many difficult-to-treat stable halogenated organics are also readily oxidized under these conditions. Chloroform and PCB (polychlorinated biphenyl) are also readily oxidized (Prengle *et al.*, 1977). However, the cost of treatment for large flows of water containing aliphatic halocarbons is relatively high due to the poor UV absorbance of those compounds.

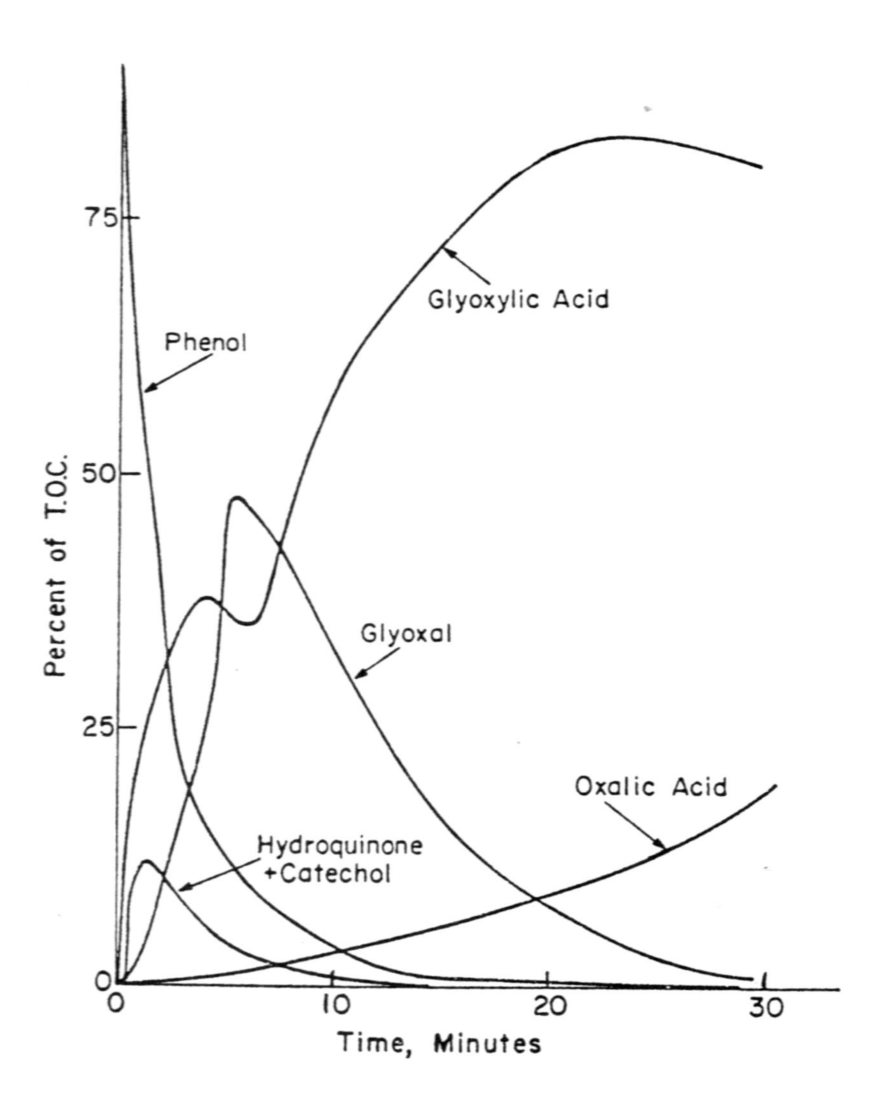

Figure 11. Variations of all intermediates as a function of time. Compounds are given as percentages of total organic carbon.

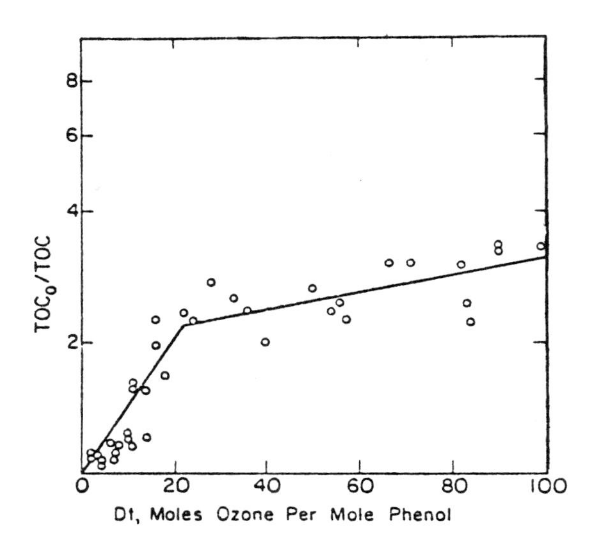

Figure 12. Total organic carbon removal as a function of the ozone dosage.

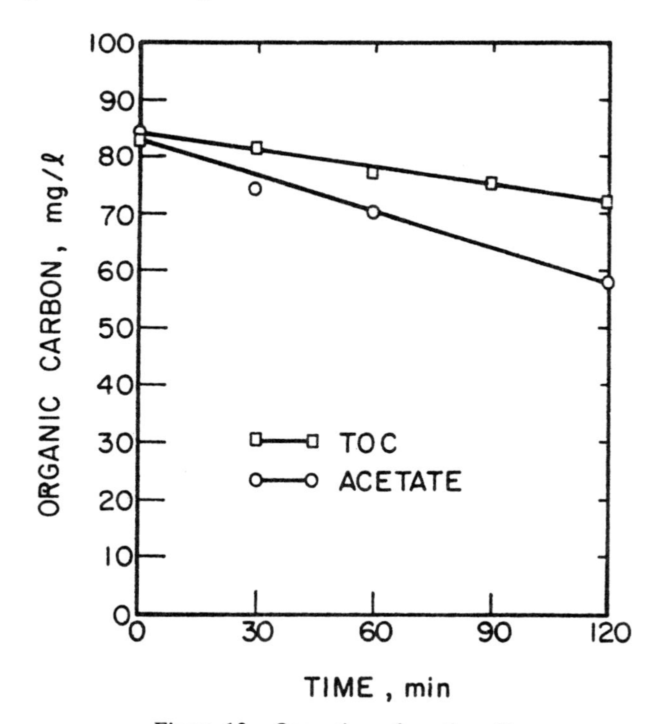

Figure 13. Ozonation of acetic acid.

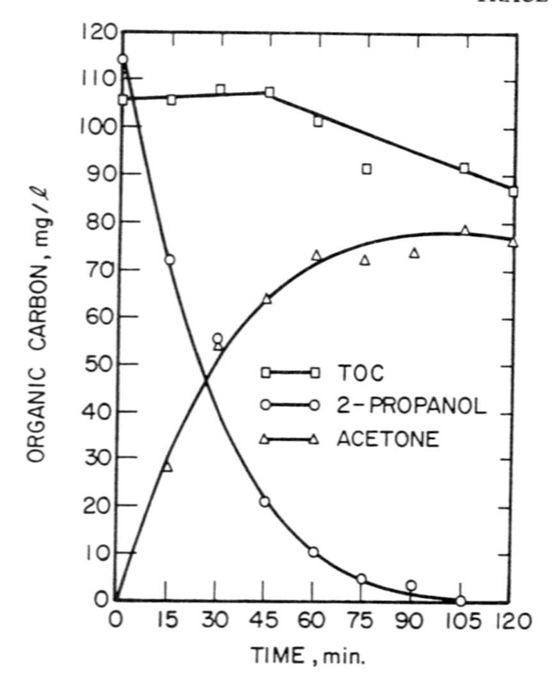

Figure 14. Ozonation of 2-propanol.

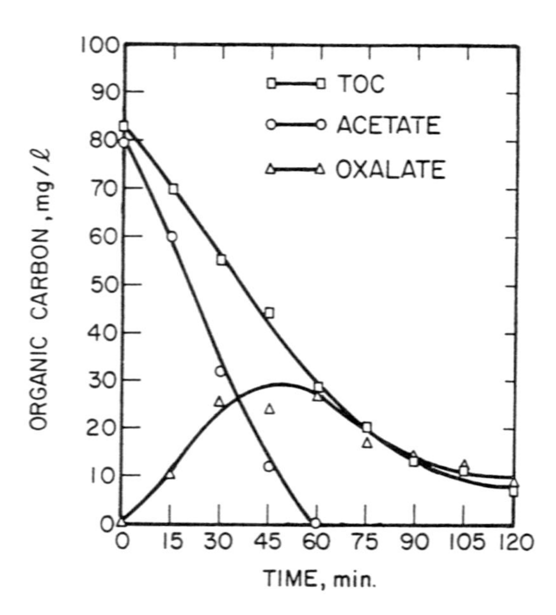

Figure 15. UV-ozonation of acetic acid.

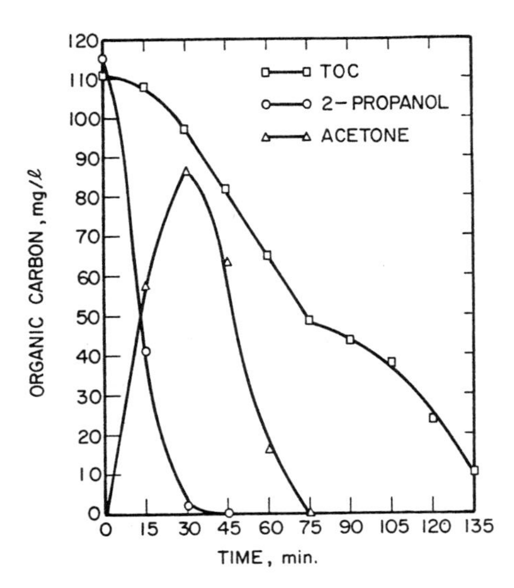

Figure 16. UV-ozonation of 2-propanol.

The halogenated organic compounds detected in finished drinking water are shown in Table II. Some of these compounds have been shown to be potentially carcinogenic (Bellar *et al.*, 1974; Stevens *et al.*, 1975; Coleman *et al.*, 1975). Their removal by ozone alone is rather slow and requires massive dosages of ozone.

Table II. Halogenated Organic Compounds in Finished Drinking Water

Compound	Formula	Range (μg/l)
Chloroform	$CHCl_3$	<0.1 – 311
Bromodichloromethane	$CHBrCl_2$	NF – 116
Dibromochloromethane	$CHBr_2Cl$	NF – 100
Bromoform	$CHBr_3$	NF – 92
1,2-Dichloroethane	$(CH_2Cl)_2$	NF – 6
Carbon Tetrachloride	CCl_4	NF – 3

NF = None Found.

The reaction of chloroform with ozone is very slow, and essentially no chloroform removal occurs with low dosages of ozone commonly used in water treatment as shown in Figure 17 (Love *et al.*, 1976; Hubbs, 1977). Although they can be essentially completely removed by very large dosages of ozone during very long contact times as shown in Figure 18, such a process is economically unacceptable.

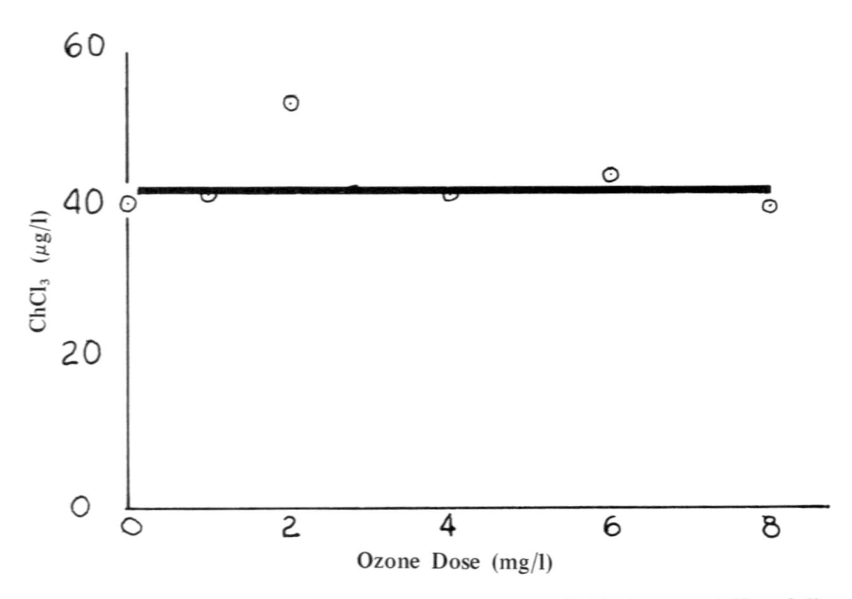

Figure 17. Chloroform oxidation vs ozone dosage, finished water (pH = 8.5).

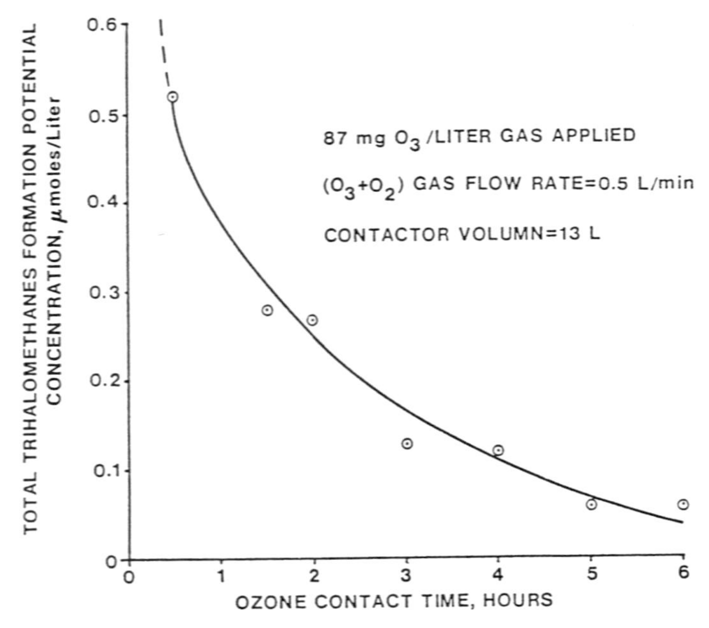

Figure 18. Batch ozonation to reduce trihalomethane formation potential in Ohio River Water.

Substantial chloroform reduction occurs when the chloroform precursors are treated with ozone prior to settling and filtration of the water as indicated by Figure 19 (Love *et al.*, 1976; Hubbs, 1977). Using this method ozone prevents the formation of chloroform by reacting with the precursor. Because of the small ozone dosage applied, only a partial oxidation can occur to more polar carboxylic acid aldehyde and ketone products. These more polar compounds will be adsorbed more strongly on the surface of polar adsorbers and more readily removed during settling.

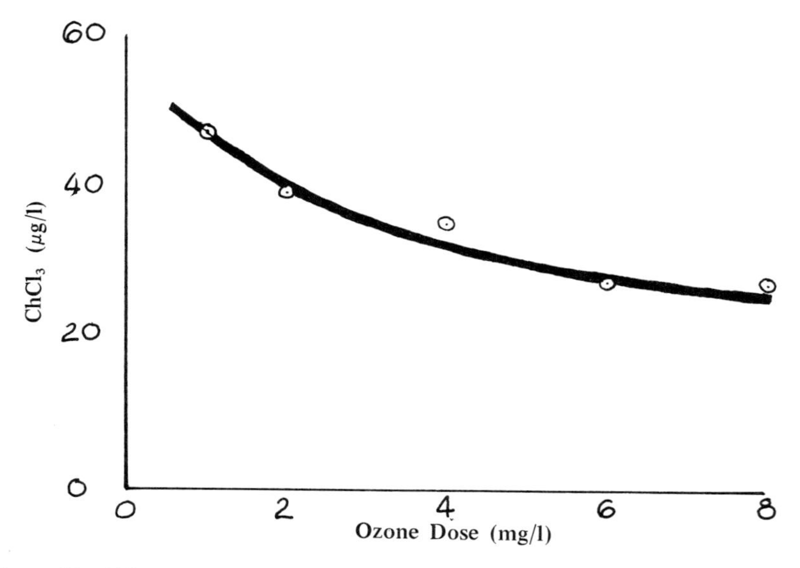

Figure 19. Chloroform precursor oxidation vs ozone dosage, presettled water (pH = 7.1).

OZONE AND ADSORPTION

LePage (1974) found during a pilot study that preozonation of the raw water much improves the alum coagulation-flocculation settling process. The floc density and time of floc formation could be correlated with the ozone dosage applied.

The same study also noted that the chlorine demand of the raw water decreased after the ozone treatment even prior to flocculation and settling as indicated by Figure 20. The chlorine demand reduction is not linearly proportional to the ozone dosage. The most probable explanation for this observation is that some of the ozone oxidation products react more rapidly with chlorine than others.

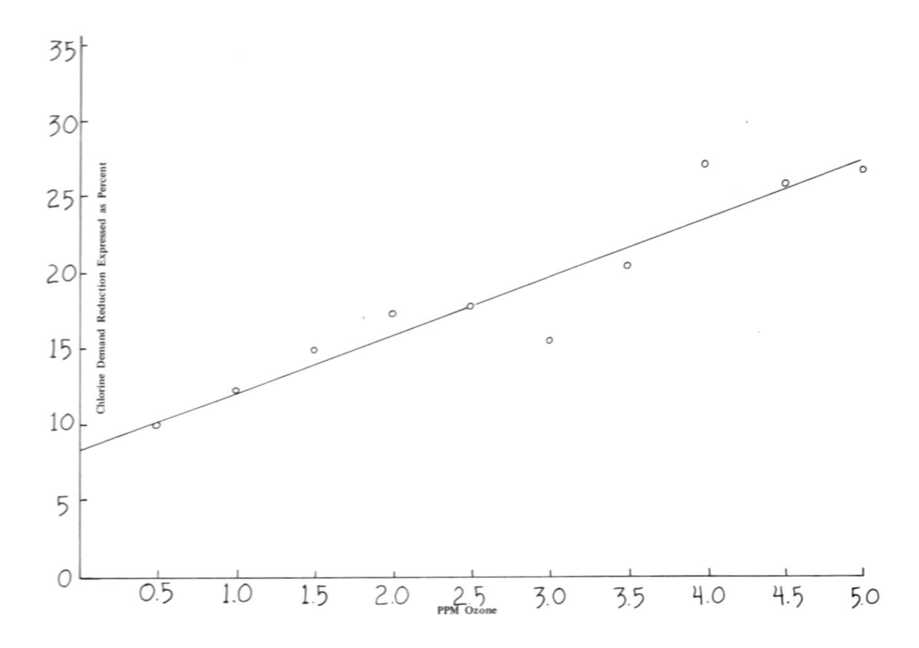

Figure 20. Example of chlorine demand reduction by ozonation.

The reduction in chloroform formations is the result of lower chlorine demand after ozone treatment as well as of more effective coagulation-flocculation settling after ozone treatment. The adsorption is dependent on the polarity of the adsorber as shown in Figure 21 (Kuhn *et al.*, 1977). Ozone treatment improves adsorption on the polar alumina but decreases the adsorption on the nonpolar activated carbon.

Increased ozone dosage leads to increased adsorption on the polar calcium carbonate adsorber as indicated by Figure 22. However, after a certain ozone dosage no further improvement occurs on aluminum sulfate. A large ozone dosage leads to less adsorption on activated carbon.

OZONE AND ACTIVATED CARBON BIOLOGICAL FILTER

The ozone literature makes numerous references to the fact that ozone treatment often increases the BOD of wastewater at first by converting non- or poorly biodegradable organics into more readily biodegradable compounds (Nebel *et al.*, 1973; Rosen, 1976; Bollyky, 1975). The renewed germ formation is much more rapid if the water is preozonated. The data shown in Figure 23 indicate that the higher the ozone dosage the faster the renewed germ formation.

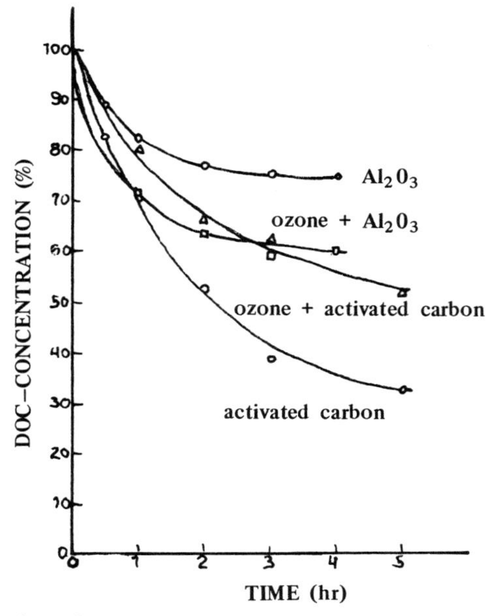

Figure 21. Adsorption of ozonized and nonozonated wastewater on Al_2O_3 and activated carbon.

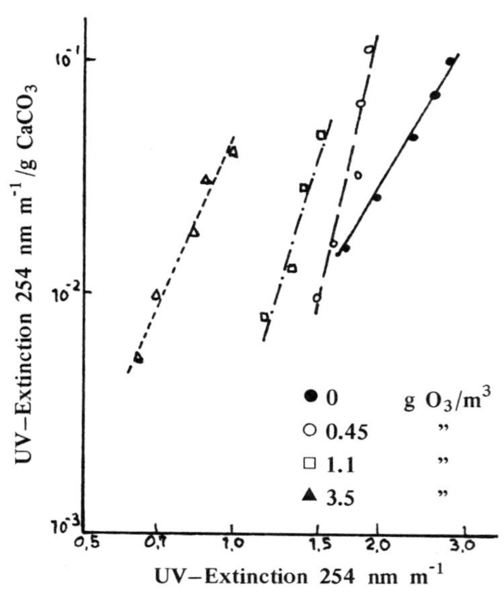

Figure 22. Adsorption of organic material from ozonated Lake of Constance water onto $CaCo_3$.

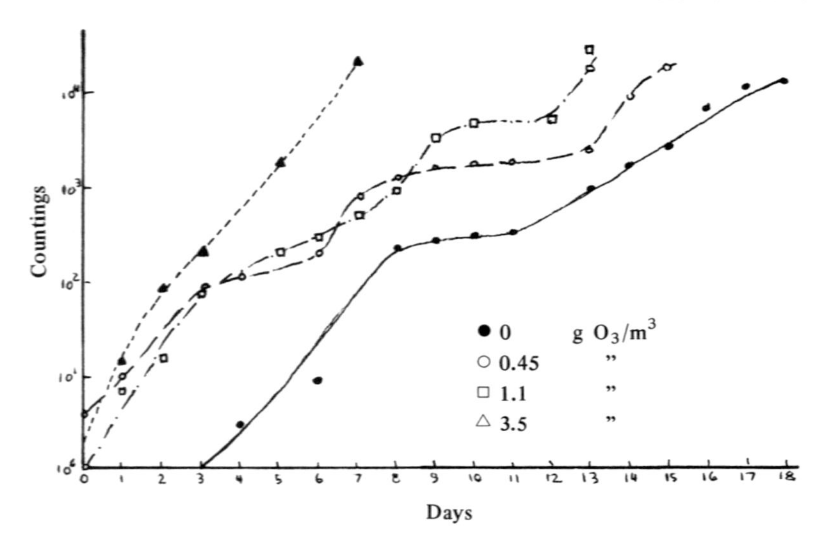

Figure 23. Renewed germ formation in Lake of Constance water after different ozone dosages.

The water treatment plants along the Rhine river in Germany often treat the river water by sand bank filtration, then by ozone and finally it is passed through activated carbon filters (Hopf, 1976; Kuhn *et al.*, 1977). It is an efficient, simple and very effective process (see Figure 24). The sand bank filtration removes many of the contaminants from the water (see Figure 25). The ozone treatment of this relatively high quality raw water is followed by filtration on activated carbon.

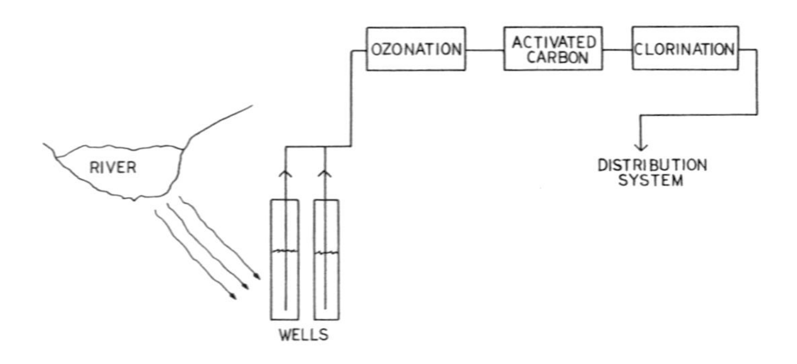

Figure 24. Inground sand filtration followed by ozone treatment and activated carbon biological filter.

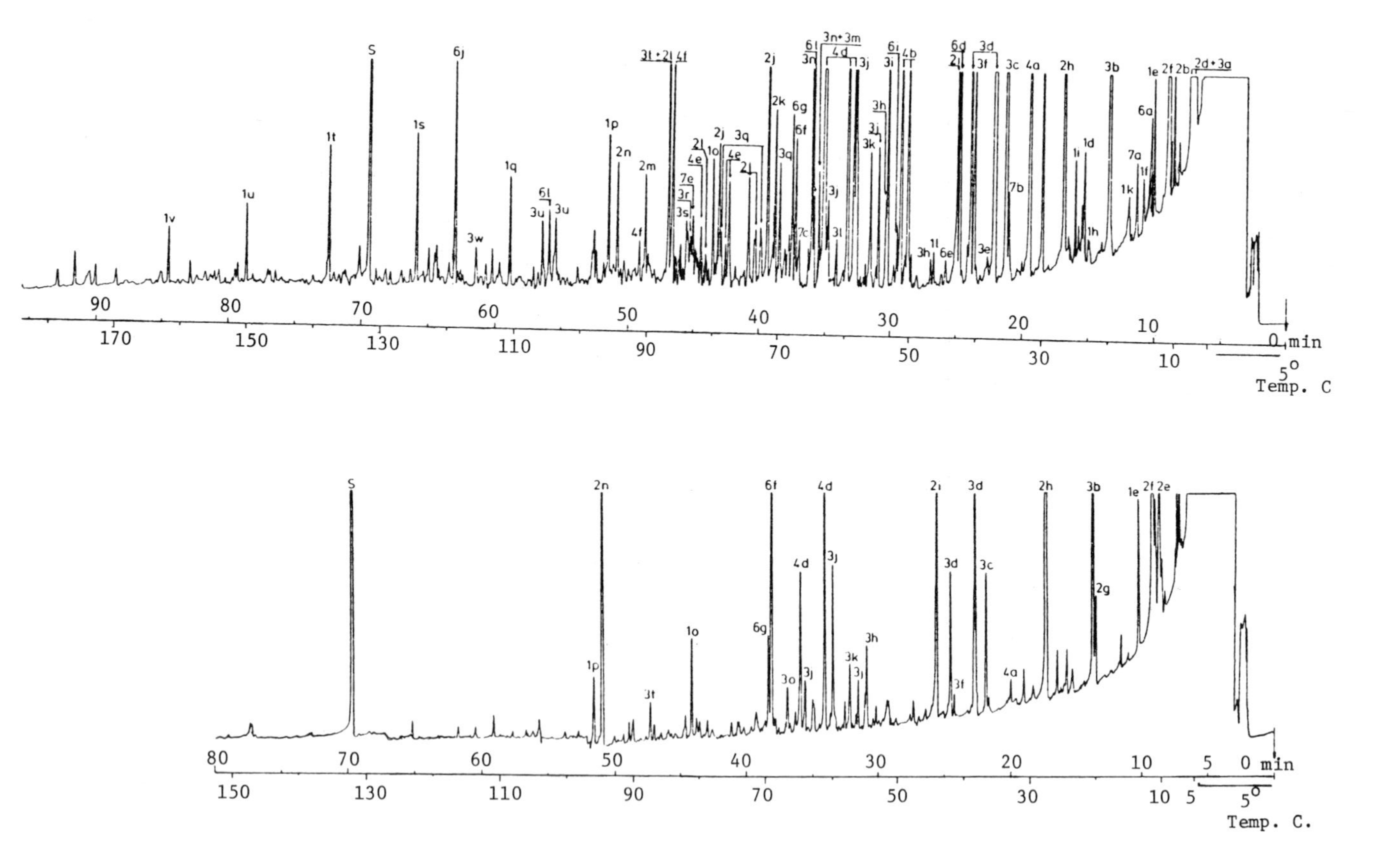

Figure 25. GC/MS chromatograms of Rhine water and sand bank filtrate.

There is strong evidence that biological activity occurs on the filter. First, the life of the carbon is much longer than expected for an exclusively adsorption-type process. Second, ozone treatment would not be expected to improve adsorption on the carbon. Therefore, another process must operate.

An improved polishing performance of activated carbon was observed by Guirguis *et al.*, (1976) after preozonation of treated municipal and industrial effluent from a physical-chemical treatment process. A superior effluent quality was obtained when ozone treatment preceded the carbon filter as compared to when it followed it. The life of the carbon was much longer when preozonation was practiced.

The process of ozone treatment followed by activated carbon filter appears to be a promising method for the treatment of raw water and for the polishing treatment of wastewater effluents. The advantages of the process are summarized in Table III. This process also renders the activated carbon treatment economically more attractive by lengthening the life of carbon. The cost of replacement carbon is a substantial item in the overall cost of activated carbon filtration as shown in Table IV (Clark *et al.*, 1976).

Table III. Treatment of Raw Water by Ozone Followed By Activated Carbon Biological Filter

Converts nonbiodegradable organics into biodegradable.
Biological activity on filter removes organics.
Effluent has low bacteria count ($>$ 10/ml).
Infrequent reactivation of carbon.
Low chlorine demand in effluent; 0.1 - 0.3 mg/l is sufficient for disinfection.
Lower cost than for the two separately.

Much more work is needed to study the mechanism of the process and to establish its limitations and potentials. The process is clearly dependent on the dissolved oxygen in the water as the source of oxygen for oxidation. This dependence alone would limit the organic load that could be treated. Therefore, the likely candidates for this treatment should have relatively low BOD and COD.

Ozone treatment is a powerful, effective process for the removal of trace organics either alone or in combination with UV irradiation or activated carbon filter. The trace organics to be removed determine the type of ozone process required. Ozone treatment alone or in combination with UV might be selected on economical grounds when the partial oxidation of the organics to harmless or less harmful partial oxidation product is

Table IV. Comparison Among Systems (¢/1000 gal)

System	1 mgd	5 mgd	10 mgd	100 mgd	150 mgd
Chlorination	3.6	1.6	1.2	0.7	0.6
Chlorine Dioxide	3.8	2.0	1.7	1.3	1.3
Ozonation (air)	6.3	2.3	1.6	0.8	0.8
Ozonation (oxygen)	7.7	2.5	1.8	0.9	0.8
Aeration	22.3	14.6	12.8	9.0	8.5
Granular Activated Carbon (media replacement)	41.0	15.5	11.7	6.0	5.1
Granular Activated Carbon (contactors)	46.3	17.5	13.6	6.7	6.3

required. Alternately, ozone treatment followed by an activated carbon filter may be the more economical choice when the complete oxidation to carbon dioxide or the complete removal of the organics is required.

REFERENCES

American Chemical Society. "Ozone Reactions with Organic Compounds," Am. Chem. Soc. Series 112 (1972).

Bellar, T. A., J. J. Lichtengerg and R. C. Kroner. "The Occurrence of Organohalides in Chlorinated Drinking Water," EPA-670/4-74-008, EPA, Cincinnati (1974).

Bollyky, L. J. *Water Quality and Treatment,* Am. Water Works Association (New York: McGraw-Hill Book Co., 1971), pp. 173-176.

Bollyky, L. J. "Disinfection, Color, Taste and Odor Control," in *Environmental Engineers Handbook,* B. Liptak, Ed. (Radnor, Pa.: Chilton Book Co., 1974).

Bollyky, L. J. "What About Ozone? Ozone Applications for Disinfection and Odor Control in Wastewater Treatment Plants," *N. J. Effluents* 8:8-10 (1975).

Bollyky, L. J. "Ozone Provides Powerful Disinfectant for Water," *Water Sew. Works* (October 1976), pp. 66-67.

Bollyky, L. J. "Ozone Disinfection of Secondary Effluents," *Water Sew. Works* (April 1977), pp. 90-92.

Bollyky, L. J., C. Balint and B. Siegel. "Ozone Treatment of Cyanides and Plating Waste on a Plant Scale," Second International Symposium on Ozone Technology, *Int. Ozone Inst. Proceedings* (Jamesville, N.Y.: Ozone Press Internationale, 1976), pp. 393-420.

Bollyky, L. J. and C. Nebel. "Mass Transfer of Ozone into Water and Wastewater," (In Press, 1977).

Clark, R. M., D. L. Guttman, J. L. Crawford and J. A. Machisko. "The Cost of Removing Chloroform and Other Trihalomethanes from Drinking Water Supplies," in *Interim Treatment Guide for the Control of Chloroform and Other Trihalomethanes,* Appendix 1, (Cincinnati, Ohio: U.S. Environmental Protection Agency, 1976).

Coleman, E., R. Lingg, R. Melton and F. Kopfler. "Occurrence of Volatile Organics in Five Drinking Water Supplies Using GCMS," First Chemical Conference of the North American Continent, Mexico City (December 1975).

Gould, J. P. and W. J. Weber. "Oxidation of Phenols by Ozone," *J. Water Poll. Control Fed.* 48:4 (1976).

Guirguis, W., T. Cooper, J. Harris and A. Ungar. "Improved Performance of of Activated Carbon by Pre-Ozonation," *Abst. Water Poll. Control Fed.*, 49th Annual Conference (October 1976).

Heist, J. A. "Ozone Oxidation of Wastewater Contaminants," *Water* Am. Inst. Chem. Eng. (1973), pp. 456-67.

Hopf, W. "Treatment of Water with Ozone and Activated Carbon (Dusseldorf)," *Wasser-Abwasser* III (2):83-92 (1970; J. M. Symons Trip Report for European Travel with Special Emphasis on Reactivation of Granular Activated Carbon, U.S. EPA, Cincinnati, Ohio (1976), p. 8010.

Hubbs, S. A. "Oxidation of Haloforms and Haloform Precursors Utilizing Ozone," Ozone/Chloride Dioxide Oxidation Products of Organic Materials Symposium, *Int. Ozone Inst. Proceedings* (In Press, 1977).

Kuhn, W., H. Sontheimer and R. Kurtz. "Use of Ozone and Chlorine in Water Works in the Federal Republic of Germany," Ozone/Chlorine Dioxide Oxidation Products of Organic Materials Symposium, Int. Ozone Inst. Proceedings (In Press, 1977).

Kuo, P. P. K., E. S. K. Chian and B. J. Chang. "Identification of End Products Resulting From Ozonation of Compounds Commonly Found in Water," Ozone/Chlorine Dioxide Oxidation Products of Organic Materials Symposium, *Int. Ozone Inst. Proceedings* (In Press, 1977).

LePage, W. L. "Ozone Treatment at Monroe, Michigan," Second International Symposium on Ozone Technology, *Int. Ozone Inst. Proceedings* (Jamesville, N.Y.: Ozone Press Internationale, 1976), pp. 198-210.

Love, A. T., J. K. Carswell, R. J. Miltner and J. M. Symons. "Treatment for the Prevention or Removal of Trihalomethanes in Drinking Water," *Interim Treatment Guide for the Control of Chloroform and Other Trihalomethanes,* Appendix 3 (Cincinnati, Ohio: U.S. Environmental Protection Agency, 1976).

Masschelein, W., G. Fransolet and J. Genot. "Techniques for Dispersing and Dissolving Ozone in Water," *Water Sew. Works* (December 1975), pp. 57-60.

Nebel, C., R. D. Gottschling, R. L. Hutchinson, T. J. McBride, D. M. Taylor, J. L. Pavoni, M. E. Tittlebaum, H. E. Spenser and M. Fleishman. "Ozone Disinfection of Industrial-Municipal Secondary Effluents," *J. Water Poll. Control Fed.* 45:2493-2507 (1973).

Oehlschlaeger, H. J. "Reactions of Ozone with Organic Compounds," Ozone/Chlorine Dioxide Oxidation Products of Organic Materials Symposium, *Int. Ozone Inst. Proceedings* (In Press, 1977).

Prengle, H. W., C. E. Mauk and J. E. Payne. "Ozone/UV Oxidation of Chlorinated Compounds in Water," Forum on Disinfection, *Int. Ozone Inst. Proceedings,* (Jamesville, N.Y.: Ozone Press Internationale, 1977). pp. 286-295.

Richard, Y. and L. Brener. "Organic Materials Produced on Ozonation of Water," Ozone/Chlorine Dioxide Oxidation Products of Organic Materials Symposium, *Int. Ozone Inst. Proceedings* (In Press, 1977).

Rosen, H. M. "Wastewater Ozonation: A Process Whose Time has Come," *Civil Eng. ASCE* (March 1976), pp. 65-69.

Stahl, E. "Ozone Contacting Systems," First International Symposium on Ozone for Water and Wastewater Treatment, *Int. Ozone Inst. Proceedings* (Jamesville, N.Y.: Ozone Press Internationale, 1975), pp. 40-45.

Stevens, A. A., C. Slocum, D. Seeger and G. Robeck. "Chlorination of Organics in Drinking Water," Conference on Environmental Impact of Water Chlorination, Oak Ridge National Laboratory (October 1975).

7

THE DETERMINATION OF CARCINOGENESIS INDUCED BY TRACE CONTAMINANTS IN POTABLE WATER

H. F. Kraybill

Division of Cancer Cause and Prevention
National Cancer Institute
Bethesda, Maryland

INTRODUCTION

An increasing awareness of the potential contribution of environmental factors to human cancer has focused much attention on the need to delineate environmental etiologic agents. This has consequently resulted in state and national legislative mandates for control of recognized and suspect carcinogenic agents in air, water and the food supply. It has been postulated that a high percentage (60-90%) of human cancers originate from environmental causes. Regardless of the debate over the percentage contributions to radiation, viruses, genetic factors, cigarette smoke, occupational exposures and multiexposure to commercial chemicals, it is readily apparent that a large percentage of human cancers could be prevented by considering the quality of our environment. National programs that appreciate the significance of environmental health ultimately lead to popular demands for remedial measures.

Modern instrumentation in the chemical laboratory has made available sophisticated methodology for detection of microlevels of contaminants, especially the volatile organics in raw water and potable water for municipalities. Various reports have appeared on this subject by many authors, but comprehensive papers by Tardiff *et al.* (1975) and Kraybill (1975, 1976a, 1977) relating to U.S. water supplies, and McNeil *et al.* (1977) on Canadian drinking water, provide a prospectus on the drinking water contaminants. In September 1976, a conference in New York City (1977) sponsored by the New York Academy of Sciences and entitled "Aquatic Pollutants and Biological Effects with Emphasis on Neoplasia," (1977) effectively coalesced all the current areas of interest with regard to identification of organic biorefractories, proposed technologies for removal, biological effects of specified biorefractories including carcinogenic and noncarcinogenic responses in test systems, and, finally, matters concerned with risk assessment and regulatory responsibilities.

The passage of the Safe Drinking Water Act of 1974 (PL 93-523) was probably accelerated by this aforementioned awareness that some of the aquatic contaminants recognized as carcinogens and those that are suspect could pose a threat to human health. Accordingly, in 1975, the National Academy of Sciences was requested by the Environmental Protection Agency, under the power of that Act, to conduct a study of the adverse health effects attributable to the carcinogenic and/or noncarcinogenic contaminants in drinking water. In this soon to be issued report, special attention was given to carcinogenic, volatile, organic biorefractories with risk assessments made wherever possible, directed toward a feasible and acceptable intake level. Feasible is indicated because there *are* limitations on engineering capabilities for removal, and an appreciation that zero levels of certain organics are unattainable and, therefore, a zero order of risk is unachievable. This report should serve as a basis for revision of the national interim drinking water regulations.

These toxic and carcinogenic contaminants occur in the parts per billion or trillion levels, and it can be expected that microexposures will occur over a long period of time. Experimental evidence on individual chemicals as to their carcinogenic potency is useful in extrapolating to human situations and potential risk, but there are limitations, which will be discussed subsequently. Similarly, epidemiological verifications are useful, but they too have their limitations insofar as spurious associations.

When one considers that there are several million chemicals in the universe (Kraybill, 1976a), those of natural origin and those from industrial technologies, it is not surprising that the pollution problem becomes difficult to control. According to Lewis (1974), it is impractical and prohibitively expensive to prove the safety or risk of each chemical compound entering the environment. Thus, new methods and approaches are needed to assess the carcinogenic potential of a whole class and mixture of chemicals as indicated by Kraybill (1977) in delineating newer approaches in this area.

ORIGIN OF CARCINOGENIC CONTAMINANTS

One liter of water completely free of organic or inorganic contaminants would be prohibitive in cost and technologically unattainable. For example, natural waters away from industrial and population centers where pollution occurs, may easily contain molecular constituents or cations and anions of geologic origin. Synthetic chemicals are dispersed into lakes and streams from atmospheric fallout. The process is reversed in that chemicals such as DDT are volatilized from bodies of water and transported through the upper atmosphere and ultimately deposited in other regions of the world. The appearance of DDT in the polar icecaps of Antarctica and in penguin body fat is evidence of migration of such organic molecules. Natural seepage of oil and polycyclic aromatic hydrocarbons (PAH) into the continental shelf and beaches poses a health problem. Wilson and coworkers (1974) have estimated the release of carcinogenic PAH into the marine environment to be at the rate of 0.2×10^6 to 6×10^6 metric tons per year. Forty percent of the world's total seepage from the ocean floor occurs in the Pacific Ocean; and the Southern California coastal area is one of the seepage-prone areas. The chlorination process, long considered an essential procedure for elimination and control of harmful pathogens, concurrently converts certain molecular species into other organic carcinogenic molecules. An illustration is the formation and escalation of levels of the suspect carcinogen, chloroform, from a few $\mu g/l$ concentration in raw water to several hundred $\mu g/l$ concentration in drinking water. The enhancement is believed due to the effect of chlorination on humic acids.

The routes by which marine pollutants are likely to enter lakes, streams and estuaries, and ultimately the sea, are through: (a) effluents from manufacturing plants, (b) municipal sewage effluent, (c) erosion or runoff from forests and lands as agricultural chemicals and soil chemicals, (d) deliberate dumping from ships and untreated sewage barged to sea, (e) offshore oil drilling operations, (f) natural oil seepage from the ocean floor, and (g) the transfer process from atmosphere to water and the reverse process with recyclic deposition in areas quite distant from the transfer process.

Delineation of Organic and Inorganic Carcinogenic Contaminants in Water Supplies

The EPA has conducted continuing surveys on raw and finished water. From March 1 to July 1, 1975, almost 700 paired samples of raw and finished water were randomly selected at each site throughout the country and analyzed for volatile organics by gas-liquid chromatography and mass spectrometry. Not all the organic chemicals identified have been quantified, but the levels of those quantitatively analyzed range from fractions of a microgram to as high as 311 μg/l for the city of Miami, Florida. More than 300 biorefractories have been identified. The identification of the biorefractories is from an EPA listing of July 1, 1976 (1976) and the levels reported in water are from an EPA report of 1975 (1975). The classification is that developed by Kraybill (1975).

Table I presents some typical recognized and suspect carcinogens and their level of occurrence in drinking water supplies in the United States.

Table I. Some Recognized and Suspect Carcinogens Identified and Their Concentrations in Drinking Water in the United States (μg/l)

Chemical	Concentration	Chemical	Concentration
Aldrin	5.4	Dieldrin	8.0
Benzene	50	DDT	–
Benzo (A) pyrene (R)[a]	0.0002-0.002	DDE	0.05
Bis (2-chloroethyl) ether	0.42	1,4-Dioxane	1.0
BHC (lindane)	–	Endrin	0.004
Carbon tetrachloride	2.0-3.0	Heptachlor	–
Chlordane	0.1	Trichloroethylene	–
Chloroform	0.1-311	Vinyl chloride (R)[a]	10.0
1,2-Dibromoethane	–		

[a]R = Recognized carcinogen

This classification of recognized and suspect is an arbitrary one developed by the author. Those in the recognized class are usually those for which there is either good epidemiological evidence, or the experimental data available reflects a broad spectrum of carcinogenic responses in various animal species and strains. It is worth emphasizing that the identification and classification has been developed, as indicated above, from epidemiological surveys recorded in the literature or through reports in the literature on the bioassay of these chemicals in experimental animals. The identification for carcinogenic activity is not through bioassay of these chemicals at the same levels that they occur in drinking water. One exception is that 1,4-dioxane was experimentally determined as a carcinogen in laboratory drinking water when the chemical was administered to rats at a level of 0.75-1.8% (Hock-Ligetti *et al.,* 1970). The bioassay level in earlier studies with rodents to test the carcinogenicity of 1,4-dioxane would, of course, be many orders of magnitude (1×10^7) above the 1 μg/l level found in municipal water supplies (Table I).

The identification of biorefractories in drinking water in the Netherlands, Canada, Czechoslovakia and other countires, has been reported by Kraybill (1977) and, accordingly, indicates their classification as carcinogens. Since only the volatile organic compounds have been identified through analysis, and only a fraction of those volatiles thus far characterized, a much larger component of the nonvolatiles in drinking water remains to be identified or quantified. Thus, one may only be seeing the tip of the iceberg, insofar as molecular organics. Most of the biological assessment on water contaminants has concentrated on the organic biorefractories. Data on inorganic carcinogenic contaminants have been reported by various investigators. However, extensive identification, monitoring and quantification of the inorganic carcinogenic compounds in raw water and drinking water must be realized so that the carcinogenic insult from these types of chemicals can be appropriately assessed. Bouquiaux (1974) has reported on the concentration of a few inorganic contaminants found in city water in The Netherlands and Germany. These typical inorganic chemicals and their levels of occurrence are shown in Table II. It is to be expected

Table II. Inorganic Suspect Carcinogens as Contaminants and their Concentrations in City Water Supplies in The Netherlands and Germany[a]

Chemical	Location	Concentration (μg/l)
Arsenic[b]	Lindau	6.2
	Dusseldorf	2.4
	Mainz	8.1
	The Hague	1.0
Cadmium	Lindau	2.0
	Mainz	9.0
Chromium[b]	Lindau	10.0
	Dusseldorf	4.5
	Mainz	9.0
Selenium	Lindau	4.9
	Dusseldorf	6.0
	Mainz	3.1

[a]From Bouquiaux (1974).
[b]Valence state unspecified.

that such inorganic chemicals occur elsewhere at the concentrations reported for these two countries. Asbestiform materials, or asbestos particles, have been found in many river systems at concentrations in excess of 10 μg/gal (Cook *et al.,* 1974; Nicholson and Farger, 1973). There is concern about asbestos since its carcinogenic activity as inhaled particles has been well-demonstrated in studies on man. However, the carcinogenic capacity of asbestos in its multiple mineralogical forms, as ingested in foods, beverages and water, remains to be assessed. The Lake Superior problem resulted from taconite tailings, containing mostly amphiboles (cummingtonite-grunerite) being dumped into the lake, which supplied water to Duluth, Minnesota. There was some chrysotile in western Lake Superior at levels of less than 1 x 10^6 fibers per liter (Cook *et al.,* 1974). Occupational exposure to inhaled asbestos fibers induces mesotheliomas and bronchogenic carbinoma in man. Asbestos had been used in filters for filtration of water, pharmaceutical solutions, etc. This use probably accounts for its occurrence in beverages and food products.

In some locations near asbestos mines the levels of asbestos fibers in waterways are quite high. For example, in the San Francisco Marin County area of California the counts approached 2 x 10^6 fibers per liter (Robeck, 1974). Andrew (1973), however, reported that 31 streams in Minnesota and Wisconsin had no detectable concentration of amphiboles. Kraybill (1976) has given a more detailed report on the asbestos fiber distribution in waters in various locations in the United States.

Berg and Burbank (1972) conducted a study on cancer mortality in certain river basins in the U.S. oriented around the occurrence of trace metals in the water supplies. From 16 basins, 10 were included in the study comprising entire states that were well sampled. A summary statistic on these 10 basins (the product of the frequency of detection and the average detection concentration of the metal) was calculated. The ranges of metal concentrations in the 10 river basins are shown in Table III.

Table III. Ranges of Metal Concentrations in Ten River Basins[a]

Metal	Low Frequency of Detection (%)	Basin Mean Positive Level (μg/l)	High Frequency of Detection (%)	Basin Mean Positive Level (μg/l)
Arsenic	2	53	9	68
Beryllium	0	—	14	0.3
Cadmium	0	—	8	3
Chromium	5	17	56	14
Cobalt	1	1	10	19
Iron	79	19	99	120
Lead	12	8	24	33
Nickel	2	5	25	31

[a]From Berg and Burbank (1972).

These investigators did find some correlations between levels of some of the carcinogenic metals and cancer mortalities in a geographic region. Iron, cobalt and chromium showed no significant correlations for man. However, nickel concentrations appeared to correspond with mouth and intestinal cancer death rates as did arsenic with eye, larynx and myeloid leukemia. Beryllium correlated with bone cancer as well as mortalities from breast and uterine cancer. Lead was associated with kidney cancer, leukemias, lymphomas, stomach, intestinal and ovarian cancers.

Previous reports have been given on the association of skin, lung and liver cancers in man where arsenic was ingested in the potable water supplies of Taiwan and Argentina (Kraybill, 1976b). The arsenic level in some waters are as high as 0.6 ppm, or estimates as high as 2000 μg/day for the Taiwanese water supply. There has been some speculation whether this is a direct effect of arsenic or a cocarcinogenic effect.

One cannot generalize on the carcinogenicity of these elements and compounds. For example, at low concentrations, selenium and arsenic serve a role in the function of the hematopoietic system. Chromium in the hexavalent state is toxic and carcinogenic; whereas in the trivalent state it is important physiologically as is the chromium insulin complex or the glucose tolerance factor.

Some Epidemiological Considerations

More than 300 biorefractories (organic) have been identified in drinking water. As previously indicated, this may be a small component of the total contaminants since there are still unidentified volatiles, some non-volatiles and of course, some inorganic chemicals. A recent EPA listing comprises some 1258 organic chemicals in 33 different types of water, raw or finished, and the frequency of their occurrence (Shackelford and Keith, 1976). A few of those identified have been shown to be toxic, that is carcinogenic, mutagenic or teratogenic when tested at high doses in animals, but sufficient data do not exist to define the effect of low-level human exposure. One study focused on the parishes of Louisiana because of the suspicions of a possible etiologic role of drinking water in the high incidence of bladder cancer in New Orleans (Page *et al.,* 1976). Multivariant regression analysis indicates a statistically significant relation between cancer mortality rates in Louisiana and drinking water obtained from the Mississippi River. This relationship would appear to hold for cancer of the urinary organs and gastrointestinal tract. There are some caveats to be attached to this study in that the effect of occupational exposures could be buried in the aggregate data. Furthermore, cancer rates in New Orleans could be inflated by movement of people from elsewhere into the area, as well as regional effects of diet and the possible contribution of air pollutants. Some data on tobacco and alcohol consumption and air pollution by parishes, were unavailable. While there may appear to be an association between carcinogens in drinking water and cancer mortality, this statistical study alone cannot prove causality.

An unpublished report by Buncher (1976) proposes that cancer incidence in the vicinity of Cincinnati in a population ingesting ground water differs from that reported for those obtaining surface water. We have been unable to evaluate this study at this time.

Another study, conducted by Cantor and associates (1977) approaches the problem by relating cancer mortality in geographic regions to the distribution of certain environmental carcinogens. This study deals with specific cancer mortality rates by site, sex and race compared to levels of drinking water contaminants in 76 U.S. counties. The data on total trihalomethanes is separated into chloroform and nonchloroform components. Chloroform is highly correlated with total trihalomethanes since it contains the major component of the trihalomethane fraction. Some of the brominated compounds, such as bromoform, dibromochloromethane and dichlorobromomethane, mutagenic in the Ames test system (based on use of bacterial strains), when used as an exposure index in geographic correlation studies appeared to show a stronger association than chloroform with certain cancer mortality rates, especially bladder cancer in males and females. These brominated alkanes have not been extensively bioassayed in experimental animals for their carcinogenic activity.

A summary by one British reviewer (Bibra, 1976) indicates that "one study of 50 water supplies has indicated a statistically significant correlation between the number of deaths due to cancer in 1969-71 and chloroform concentration in the spring of 1975, but no such correlation was found with the total number of deaths or with concentrations of other trihalomethanes. Another similar epidemiological analysis of 43 cities failed to confirm the findings in the first study. Furthermore, what is missing is some data on the contribution of drinking water to the total human exposure to halocarbons. The contribution from water could be small or certainly influenced by the variant of halomethanes or haloalkanes in food and air." More definitive data are needed to assess the shortcomings of the various studies, or conversely, to establish that there are no spurious associations.

The previously mentioned report by Buncher (1976) indicates there was a higher risk of cancer (liver, pancreas and bladder) for those members of the population who have been exposed to surface water as their source of drinking water compared with those using ground water. The risk was 5-7% higher in males and 2-3% higher in females. It is worth emphasizing that these findings contradict those of an earlier, less-extensive study. We have been unable to critically evaluate this report. While some of these studies omit variables and may be viewed by some as spurious associations, and certainly not proof of causality, they do provide some leads for further approaches to verification. In this regard, the National Cancer Institute is considering demonstration projects that would provide for more definitive studies, buttressed by etiological and epizootic findings, that would provide more substantiation of causal associations in future epidemiological pursuits.

Other Approaches to Assessment and Confirmation of Carcinogenicity

Municipal potable water and raw water are composed of many chemical contaminants. In drinking water, man receives this composite of organic and inorganic micropollutants. Tardiff and Deinzer (1973) have reported on the preparation of a concentrated extractable residue, which can be isolated and which contains proportionate amounts of readily identifiable components. Thus, water sample fractions and concentrates become available for toxicological or biological assay. Through reverse osmosis or ultrafiltration, osmotic ultrafiltration, evaporation and lyophilization, a freeze-dried concentrate of the order of 2 g can be reclaimed from the processing of 500 gallons of water. Obviously, such small gram fractions cannot be checked for carcinogenic activity by the traditional *in vivo* bioassay of feeding an animal the test chemical(s) over a two-year period. One therefore must resort to procedures using cell cultures for measurement of *in vivo* or *in vitro* cell transformation as delineated by DiPaolo *et al.*(1973) or by means of single injections to newborn mice (Kelly and O'Gara, 1973).

A new test that screens for chemical carcinogens has also been developed by research scientists at the Upjohn Company in Kalamazoo, Michigan (1976). In this procedure, Chinese hamster lung fibroblasts are cultured with radioactive thymidine, grown in a nonradioactive medium, and finally exposed for one to four hours to the test substance in the presence of liver cells called microsomes. The microsomes are added to simulate metabolic processes, which render otherwise inactive agents carcinogenic. The extent of DNA damaged by the test chemical is determined by measuring the intact and filterable radioactive DNA after treated cells are disrupted and subjected to alkali. Specific biorefractories or a composite of biorefractories could be screened for carcinogenicity using this rapid procedure.

Mutagenic activity is closely correlated with carcinogenic activity. In a screening of about 165 chemicals in the United States and Japan by a battery of assay systems, DeSerres (1975) reported that in these assays, at least 80% of the chemical carcinogens are mutagens and less than 10% of the chemicals believed to be noncarcinogenic gave false positive results. The nonmammalian microbial assay systems (*S. typhimurium, E. coli* and *B. subtilis*) are used to detect genetic alterations. By the mutagenic activity, many of the halogenated organics that are biorefractories in water have been checked for mutagenicity. Quite a few of the halogenated alkanes have been shown to be mutagenic. This prescreen information is used to selectively assay these chemicals for carcinogenic activity by the standard bioassay.

These rapid *in vitro* assays lend themselves to the determination of the mutagenic activity of solvent fractionated material from composites of water biorefractories. The specific activity of certain fractions thus serves to characterize what combinations of biorefractories represent the predominant activity, as contrasted with those of no or little activity. Since the cell culture and bacterial systems can provide a direct measure of mutagenic activity, and *in vitro* metabolic activation provides a measure of the activity of metabolites, a spectrum of significant information can thus be accrued.

Indirect and direct approaches have been used to characterize carcinogenic contaminants in water. The direct approach previously described is to monitor their presence in water. An indirect approach is to record the level and presence of chemical carcinogens in a water supply on the basis that such chemicals were proven carcinogenic in experimental animals. A study involving the direct approach using fish as the test animal is reported by Brown *et al.* (1973). The significant finding in those studies is that they observed a higher frequency of tumors in fish in polluted waters of the Fox River Watershed than in fish in relatively nonpolluted waters. The incidence of tumors found in 2121 fish examined from the polluted watershed was 4.48% compared with a tumor incidence of 1.03% in 4369 fish taken from nonpolluted waters. Tabulations on chemical pollutants are shown in Table IV.

These findings in fish suggest the utility of the fish as an animal test system for individual water biorefractories, concentrates of biorefractories and fractions of concentrates of biorefractories in a laboratory test situation. The chemical can be introduced into the water and after exposure to the fish eggs, which hatch and develop into fry and ultimately adults, the tumor response can be recorded in thousands of marine animals, making this a most suitable test procedure. One can thus appreciate that observations on feral populations can be made and lead to two classes of information; first, the wholesomeness of water using the fish as a monitor, and second, the tumorigenic response in fish from water where the chemical profile has been obtained by suitable chemical analysis using gas-liquid chromatography and mass spectrometry. Beyond analysis of sediment and fish (body burden), these tests should provide ancillary data to correlate with tumor incidence. Progressing further with this ecological approach, one can correlate such data from the marine animals and the

Table IV. A Partial Listing of Organic Compounds and Mixtures Found in Fox River Watershed[a]

Chemical Compound or Mixture	Persistence	Concentration (ppm)
Ethyl ether	1-2 years	0.002
Crude oil[b]	1-2 years	0.01
Gasoline	6 months	0.01
Benzanthracene[c]	1-2 years	0.01
Organophosphates	7-84 days	0.01
Toluene	1-2 years	0.01
Benzene[b]	1-2 years	0.1
Naphthalene	1-2 years	0.1
Benzoic Acids	3-12 months	0.1
Triazines[b]	6 months	0.2
Toluidine	6 months	0.2
Chlorinated Hydrocarbons[b]	2-5 years	0.2

[a]Adapted from Brown *et al.* (1973).
[b]Recognized and suspected carcinogens–only some triazines and chlorinated hydrocarbons.
[c]Weakly carcinogenic.

aquatic environment as presumptive evidence to investigate human cancer mortality experiences. Therefore, all these data bases should be fully exploited to enhance the association of types and levels of water contaminants with incidence of human cancer in various regions. A schematic representation of these data resources and their interrelationships in establishing any causality is shown in Figure 1.

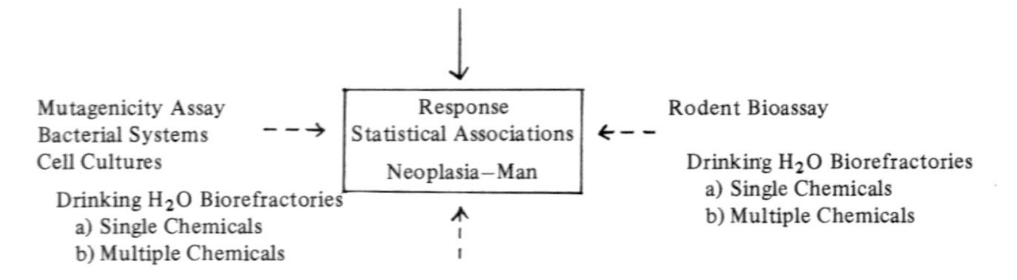

Figure 1. Relationship of water contaminant levels to incidence of human cancer.

SUMMARY

Many of the contaminants in raw water from rivers, estuaries, lakes and streams have been identified and quantified. The biorefractories in potable or drinking water on a regional basis and by municipality have also been monitored. This identification and quantification is a continuing process. Most of the effort has been expended on monitoring the organic chemical contaminants. For the organic chemical biorefractories, the current analytical procedures, using gas-liquid chromatography and mass spectrometry, have been employed to report volatile type organics. Quantitatively and qualitatively, there is a large component of non-volatiles, unidentified volatiles and inorganic chemicals present that need to be identified. Any carcinogenic insult to man could be the result of an insult from the organic biorefractories, the inorganic contaminants or a combination of all these constituent etiologic agents.

While the monitoring effort on water systems in the United States has recently been accelerated by the Environmental Protection Agency, more extensive monitoring will still be required to develop systematic studies on a geographical basis to ferret out causal relationships, if any, from man's exposure to drinking water. To our knowledge, only a few epidemiological studies have been conducted dealing with various statistical associations between cancer mortality incidence, types of drinking water (surface and ground) and types of inorganic and organic chemicals. Such a study may show some positive associations. Some of the findings may be tentative conditioned by the influence of other environmental variants. Some investigators have recognized the need to acquire ancillary data, which would enhance the validity of the observations to human neoplasia and association of the aquatic variants.

The significance of other evidence, such as ecological information and epizootic data, is mentioned as an additional approach in pursuing such complex studies. Environmentally, with respect to the total contaminated insult man may receive, the fractional contribution from contaminanted water and fish that reflect this contamination and are consumed by man may not rank as high as the environmental load from diet contaminants and air pollutants. For this and other reasons the carcinogenicity studies that apply to humans must withstand a critical evaluation as to design and validity of any conclusions developed.

Some of the organic biorefractories in drinking water have been identified as carcinogens through references to reports that some of these chemicals have been bioassayed in standard procedures using the rodent and/or other species. Similarly, some of the inorganic contaminants in water have been classified as carcinogens from experimental studies.

Some epidemiological studies have been conducted on inorganic contaminants; these attempt to show, on a regional basis, some possible relationships to cancers in man in various organs and tissues. Some of these studies raise questions that may necessitate further investigations.

In addition to the corroborative evidence from carcinogenicity bioassay on the various organic chemicals, using typical rodent feeding and inhalation studies over a two-year period, other approaches are suggested for determination of carcinogenicity. The use of bacterial systems and cell cultures is recommended to determine mutagenicity and cell transformation in the *in vitro* bioassay, to augment *in vivo* procedures.

Assessment of the role of aquatic carcinogens in the potential induction of human cancer is a complex area for research that requires the contributions and achievements of a multidisciplinary approach. Progress is being made because water as an environmental stress system is perhaps better identified, classified and monitored than any other human exposure. While the carcinogenic stress may not be dependent on a microlevel to any one molecular species, there is the inevitable possibility that the total dose from multiple biorefractories and/or contaminants may be adequate to initiate the response over a latent period of many years.

REFERENCES

Andrew, R. W. "Mineralogical and Suspended Solids Measurements of Water Sediment and Substrate Samples for 1972," Lake Superior Study II, Stream Sediments Data Report (April 1973).

Berg, J. W. and F. Burbank. "Correlations Between Carcinogenic Trace Metals in Water Supplies and Cancer Mortality," *Ann. N.Y. Acad. Sci.* 199:249-264,(1972).

Bouquiaux, J. "Nonorganic Micropollutants of the Environment Prepared for the Commission of the European Communities," V/F/1966/74E, Luxembourg (1974).

British Industrial Biological Research Association (BIBRA). "How to Control Organic Compounds in Drinking Water," Information Bulletin-BIBRA 15(8):419 (1976).

Brown, E. R., J. J. Hazdra, L. Keith, I. Greenspan, J. B. G. Kwapinski and P. Beamer. "Frequency of Fish Tumors Found in Polluted Watershed as Compared to Nonpolluted Canadian Waters," *Cancer Res.* 33: 189-198 (1973).

Buncher, C. R. "Ohio Study Links Cancer to Contaminated Drinking Water," *Toxic Mat. News* 3 (17):133 (August 18, 1976).

Cantor, K. *Proceedings of Conference on Aquatic Pollutants and Biological Effects with Emphasis on Neoplasia* September 27-29, 1976, in *Ann. N.Y. Acad. Sci.* (In press).

Cook, P. M., G. E. Glass and J. H. Tucker. "Asbestiform Amphibole Minerals: Detection and Measurement of High Concentration in Municipal Water Supplies," *Science* 185:853-855 (September 6, 1974).

DeSerres, F. J. "The Utility of Short-term Tests for Mutagenicity as Predictive Tests for Carcinogenic Activity," *Excerpta Med.* International Congress Series No. 376; and *Proc. Europ. Soc. Toxicol.* 17:113-116 (1975).

DiPaolo, J. A. and R. L. Nelson. "Host Mediated *in vivo in vitro* Assay in Chemical Carcinogenesis," *Arch. Pathol.* 95:380-385 (1973).

Hoch-Ligetti, C., M. F. Argus and J. C. Arcos. "Induction of Carcinomas in the Nasal Cavity of Rats by Dioxane," *Brit. J. Cancer* 24:164-167 (1970).

Kelly, M. and R. O'Gara. "Induction of Tumors in Newborn Mice with Dibenz (a,h) Anthracene and 2-methylcholanthrene," *J. Natl. Cancer Inst.* 26 (3):651-679 (1973).

Kraybill, H. F. "Origin, Classification and Distribution of Chemicals in Drinking Water with an Assessment of their Carcinogenic Potential," *Proceedings of the Conference on Environmental Impact of Water Chlorination,* October 22-24, 1975, Oak Ridge National Laboratory, Oak Ridge, Tennessee, Report Conf-751096 (1975).

Kraybill, H. F. "The Distribution of Chemical Carcinogens in Aquatic Environments," *Proceedings of UICC Symposium,* Cork, Ireland. *Prog. Exptl. Tumor Res.,* F. Homberger, Ed. Vol. 20 (Basel, Switzerland: S. Karger, 1976a), pp. 3,4.

Kraybill, H. F. "Carcinogenicity of Arsenic: Experimental Studies," *Proceedings of Symposium on Health Effects of Occupational Lead and Arsenic Exposure,* B. W. Carnow, Ed. HEW Publ. No. (NIOSH) 76-134 (Washington, D. C.: U.S. Government Printing Office, 1976b) pp. 272-283.

Kraybill, H. F. "Global Distribution of Carcinogenic Pollutants in Water," *Proceedings of the New York Academy of Sciences Conference on Aquatic Pollutants and Biological Effects with Emphasis on Neoplasia,* New York, September 27-29, 1976 (In press).

Kraybill, H. F. "Conceptual Approaches in Non-occupational Carcinogenesis," in *Advances in Toxicology: Environmental Cancer,* H. F. Kraybill and M. A. Hehlman, Eds. (Washington, D. C.: Hemisphere Publ., In press).

Lewis, R. "Environmental Search for the Source of Cancer," *SR/World - Science Section:* 50-51 (April 20, 1974).

McNeil, E. E., R. Otson, W. F. Miles and J. M. Rajaballe. "Occurrence of Chlorinated Pesticides in Ottawa Drinking Water," (In press).

New York Academy of Sciences. *Annals of Conference on Aquatic Pollutants and Biological Effects with Emphasis on Neoplasia,* September 27-29, 1976, H. F. Kraybill, R. G. Tardiff, C. J. Dawe, and J. C. Harshbarger, Eds. (In press).

Nicholson, W. J. and A. M. Farger. "Environmental Contamination Sources," Meeting on Biological Effects of Asbestos, National Institutes of Health, Bethesda, Maryland, February 1, 1973 (unpublished).

Page, T., R. H. Harris and S. S. Epstein. "Drinking Water and Cancer Mortality in Louisiana," *Science* 193:55-57 (1976).

Robeck, G. Personal communication from the Environmental Protection Agency Laboratories in Cincinnati, Ohio (1974).

Shackelford, W. M. and L. H. Keith. "Frequency of Organic Compounds Identified in Water," Report from Environmental Research Laboratory, EPA, Athens, Georgia (1976).

Tardiff, R. G. and M. Deinzer. "Toxicity of Organic Compounds in Drinking Water," presented at 15th Water Quality Conference, February 7-8, 1973, University of Illinois, Urbana (Unpublished).

Tardiff, R. G., G. F. Craun, L. J. McCabe and P. E. Bertozzi. "Suspect Carcinogens in Water Supplies," Interim Report to Congress, Appendix B, Health effects caused by exposure to contaminants. Environmental Protection Agency Progress Report, Cincinnati, Ohio (April 1975).

United States Environmental Protection Agency Office of Toxic Substances. "Report to Congress on Preliminary Assessment of Suspected Carcinogens in Drinking Water," Table II: 2-7, Washington, D. C. (1975).

United States Environmental Protection Agency, Health Effects Laboratory. "Organic Compounds Identified in Drinking Water in the United States," Government distribution list as of July 1, 1976, Cincinnati, Ohio (1976).

Upjohn Company. "Mammalian Cells are Used in a New Test that Screens for Chemical Carcinogens," *Chem. Eng. News* 6:18 (December 1976).

Wilson, R. D., P. H. Monoghan, A. Osanik, L. C. Price and M. A. Rogers. "Natural Marine Oil Seepage," *Science* 184(4139):857-865 (1974).

8

EFFECTIVENESS OF ACTIVATED CARBON FOR REMOVAL OF VOLATILE HALOGENATED HYDROCARBONS FROM DRINKING WATER

Walter J. Weber, Jr.
Professor of Environmental and Water Resources Engineering
Chairman, Water Resources Program
University of Michigan
Ann Arbor, Michigan

Massoud Pirbazari and Mark Herbert
Research Associate
Environmental and Water Resources Engineering
University of Michigan
Ann Arbor, Michigan

Richard Thompson
Quality Control Officer
Arkansas Department of Pollution Control and Biology
Little Rock, Arkansas

INTRODUCTION

In July 1974, the City of New Orleans and the State of Louisiana requested that the Environmental Protection Agency undertake a survey to identify and quantitate trace organic compounds present in the New Orleans finished drinking water supply. In November 1974, the Environmental Defense Fund reported the results of a study that showed a possible link between certain cancers and consumption of Mississippi River water by Louisiana residents. Immediately the EPA confirmed the presence of suspected carcinogenic organic chemicals in New Orleans drinking water (USEPA, 1974).

Following disclosure of the existence of trace quantities of organic chemicals in New Orleans water, the EPA undertook a comprehensive nationwide survey (National Organic Reconnaissance Survey – NORS) to determine the concentrations, and potential effects, of certain organic chemicals in drinking water. The NORS program provided for analysis of 80 water supplies representing a wide variety of raw water sources, treatment techniques and geographical locations.

It has been postulated that trihalomethanes are ubiquitous in chlorinated drinking waters and, indeed, result from chlorination practice (USEPA, 1975). Therefore, chloroform, bromoform, bromodichloromethane and dibromochloromethane, along with two other volatile halogenated hydrocarbons, carbon tetrachloride and 1,2-dichloroethane, were of particular interest and focus for the NORS investigation.

Rook (1974) reported the formation of such compounds during chlorination of Rhine River water, and, in a later investigation (1976), discussed the probable role of the fulvic fraction of humic substances as precursors in the production of trihalomethanes. He elaborated on the reactivity of metadihydroxy aromatic compounds, the basic building blocks of humic substances, with chlorine. Coleman *et al.* (1975), in a five-city survey, also documented the formation of halogenated organics (aliphatic and aromatic) during chlorination of drinking water.

There is presently little concrete evidence to link such compounds in water supplies directly to cancer, birth defects, cardiovascular disease or any other physiologic anomaly (Tardiff and Deinger, 1973). The etiologies of such diseases are highly complex and not sufficiently defined. Although acute or chronic toxicity may be measured in animals using toxicological techniques, extrapolation of measurements on experimental animals to effects on human beings are difficult and often uncertain. The evaluation of possible teratogenicity and mutagenicity is even more difficult and uncertain. Nonetheless, in the absence of more definitive procedures, such bioassays may be the best available means of assessing the potential toxicity of compounds present in water supplies (Ongerth *et al.,* 1973).

In this light, it is worth noting that such bioassays have indicated leukemic reactions and bladder papilloma in laboratory mice injected cutaneously and subcutaneously with carbon chloroform extracts of water (Heuper and Payne, 1963). Further, studies conducted by Dowty *et al.* (1975) have revealed the existence of five halogenated compounds in the blood plasma of persons living in New Orleans. The concentration of carbon tetrachloride in the blood plasma was found to be substantially higher than that of the drinking water, and the authors postulated that a bioaccumulation effect might be operative in humans. Most recently, the National Cancer Institute has presented evidence that chloroform may be carcinogenic (N.C.I., 1976).

Considering such observations, the presence of haloforms in drinking water must raise concern regarding public health and safety. It is incumbent upon the water industry to ensure the safety of water supplies by developing, refining and employing treatment technologies which will minimize the formation of potentially harmful organic compounds during treatment and which will remove such compounds as are formed and/or as are present in the raw water supply.

Activated carbon treatment provides one possible means–indeed, perhaps the only means–of accomplishing both objectives; that is, removal of precursor compounds as well as removal of halogenated compounds either formed in treatment or already present in the raw water. There is a substantial amount of information available regarding the efficiency of carbon for removing taste and odor compounds from drinking water, as well as a variety of organic substances from industrial and municipal wastewaters (Weber, 1972; 1974a; 1974b; 1975; Weber and Ying, 1977). There is, however, little direct information regarding its effectiveness for providing required levels of removal of volatile halogenated substances at the extremely low initial concentrations at which these compounds are typically present in drinking waters. Furthermore, there is the distinct likelihood of complex competitive adsorptions between trace haloforms and other organic constituents of water supplies, especially humic substances.

Such information as does exist is qualitative and inconclusive. For example, Stevens *et al.* (1975) reported on pilot-plant studies in which water containing humic substances was passed through granular activated carbon columns prior to chlorination; a substantial reduction in haloform formation during chlorination was observed. In evaluating granular activated carbon bed effectiveness for removing trihalomethanes from Cincinnati tap water, it was observed that chloroform and bromodichloromethane were only partially removed while dibromochloromethane was much more effectively adsorbed (Robeck, 1975).

More fundamental and quantitative information regarding the effectiveness of activated carbon in providing continuous removal of haloforms at the microgram per liter level is required for rational design of large-scale treatment facilities for water treatment applications.

To develop a firm basis for understanding the adsorption of halogenated organic compounds by activated carbon at the extremely low levels found in water supplies, and for development of rational design criteria for such applications, the Environmental and Water Resources Engineering Group of the University of Michigan initiated a related research and development program in mid-1976, under EPA sponsorship. This chapter presents some initial findings and data regarding the adsorption of several volatile halogenated compounds on activated carbon, at concentration levels pertinent for water supplies. The chapter further provides comparisons of these findings and data with those for other organic substances for which carbon is commonly used as a treatment technique.

To evaluate the effectiveness of activated carbon for removal of organic compounds, and as a first step in developing design criteria, it is essential to determine both adsorption equilibrium and rate characteristics for the system(s) of interest (Weber, 1972; Crittenden and Weber, 1976). Adsorption equilibrium relationships are usually represented in terms of adsorption isotherms, which are useful both for representing the ultimate capacity of a carbon for adsorption of an organic compound and for providing description of the functional dependence of capacity on the concentration of the compound. Rate, or kinetic, data are significant for evaluation of the time-dependent approach to equilibrium capacity, specification of contact time, and for design of continuous-flow adsorption systems.

EXPERIMENTAL

Analytical

Several analytical techniques for volatile halogenated hydrocarbons have been reported recently. Bellar and Lichtenberg (1974) used an inert gas to transfer volatile compounds from aqueous samples, sorbed the compounds from the gas phase on a porous polymer trap, and used gas chromatography with a flame ionization or microcoulometric detector for analysis. The lower limit of detection reported was 0.5 to 1 μg/l. Nicholson and Meresz (1976), employing a scandium-tritide electron capture detector, used gas chromatography by direct aqueous injection for quantification of target haloforms. The detection limit achieved by this method was $\cong$1 μg/l. Mieure (1977) used a liquid-liquid extraction technique coupled with gas chromatographic analysis; the detection limit achieved was $\cong$1 ppb. Morris and Johnson (1976) employed a static head-space method coupled with electron capture chromatography. They reported this technique to be highly sensitive for the halogenated compounds of interest.

The stripping and preadsorption technique used by Bellar and Lichtenberg was considered too time-consuming for the large number of samples involved in the present adsorption quantitation studies. In the direct aqueous injection technique described by Nicholson and Meresz, haloform precursors decompose at high injector temperatures to produce haloforms. The concentration of halogenated compounds obtained by this method is therefore the sum of the free haloform content of the sample plus the quantity of haloforms produced in the actual analysis–referred to as "total potential haloforms." The procedure was deemed unsatisfactory for the objectives of this study. The method described by Mieure was found to be rapid and relatively efficient, and was consequently selected as the analytical procedure for the adsorption rate experiments described herein. A modification of the Morris and Johnson method provided a rapid and sensitive technique which was highly compatible with the adsorption equilibrium studies.

A Varian Model 2700 gas chromatograph equipped with a scandium-tritide electron capture detector was used. The column material was 60/80-mesh Chromasorb 101, packed in a 6-ft, 1/8-in. stainless steel column. Column temperature was varied to give the best results for the particular compound being analyzed.

Temperature control was necessary for adsorption isotherm experiments. This was accomplished by placing a mechanical shaker in a walk-in incubator. A temperature of 20°C was used for all isotherm determinations.

Reagents

Analysis of Ann Arbor tap water and the distilled water available in the laboratory revealed the presence of several organic contaminants, principally chloroform and bromodichloromethane. Ann Arbor tap water was found to contain 160 μg/l of chloroform and 33 μg/l of bromodichloromethane, while the distilled tap water contained 36 μg/l and 7 μg/l of these compounds, respectively. Oxidation of the organic contaminants by refluxing for several hours in a solution of potassium permanganate and sodium hydroxide, followed by distillation, produced water of acceptable quality, but not in sufficient quantities for the study.

After some experimentation it was determined that a high-quality water free of interfering organic compounds could be prepared in sufficient quantities by passing distilled water through a Teflon®-fitted glass column containing granular activated carbon, followed by filtration through a 0.45-μ membrane filter. The water was stored in glass bottles with Teflon-lined screw caps.

Standard stock solutions were prepared by injecting microliter volumes of the compound of interest into a 1-liter volumetric flask partially filled with high-quality water. The mixture was then diluted to volume with this same water to give a concentration of 20 mg/l. Dilutions were made from the stock solution by pipetting a known quantity of stock solution into a partially filled volumetric flask and diluting to volume with high-quality water.

Stock solutions in hexane were made by injecting microliter volumes into a 100-ml volumetric flask filled with high-purity hexane. Dilutions were then made by the same technique used for the aqueous samples.

Activated Carbon

Several different commercial activated carbons were tested for suitability of use. These included the bituminous-base Filtrasorb 400 (Calgon Corporation), another bituminous activated carbon (Nuchar; Westvaco Corporation), a lignite carbon (ICI; ICI United States, Inc.), and a peat-base carbon (Norit, American Norit Co., Inc.). Adsorption rates and isotherm capacities were measured for each carbon in completely mixed batch (CMB) reactor systems. Adsorption isotherm measurements were made with both a low-molecular-weight substance (carbon tetrachloride) and a high-molecular-weight substance (humic acid). Adsorption rate measurements were made with the more critically rate-limited of these compounds, namely humic acid. Comparative adsorption isotherms are illustrated in Figures 1 and 2, and the rate data in Figure 3. It should be noted, with respect to the rate data presented in Figure 3, that the softer ICI and Norit carbons tended to degrade in particle size during mixing in the CMB reactor. Thus, because of the effect of decreased particle size on adsorption rates in batch systems (Weber and Morris, 1964), the true adsorption rates of the commercial size granular ICI and Norit carbons are likely to be substantially lower than those indicated by the comparative data given in Figure 3. It was not possible to control particle degradation with these two carbons during the course of the tests.

Repetitive air scour tests were run to evaluate the capability of each carbon to function in routine use without excessive attrition. This is essential for both packed- and expanded-bed adsorption applications if carbon loss, high head loss due to reduction of particle size, and effluent deterioration and increased disinfectant demand due to carry-over of carbon fines is to be avoided. The better adsorption rates and capacities of the Filtrasorb activated carbon, evidenced in Figures 1-3, coupled with the clear superiority of the bituminous carbons with respect to attrition resistance, led to selection of the Filtrasorb 400 for these studies. The carbon was sieved, washed, dried and dessicated prior to weighing for use in the experiments described in the following sections.

Equilibrium Studies

Equilibrium experiments were conducted in 150-ml air-tight serum vials using weighed amounts of activated carbon in 100-ml aliquots of solution. A Teflon-coated septum was crimped onto each vial with an aluminum

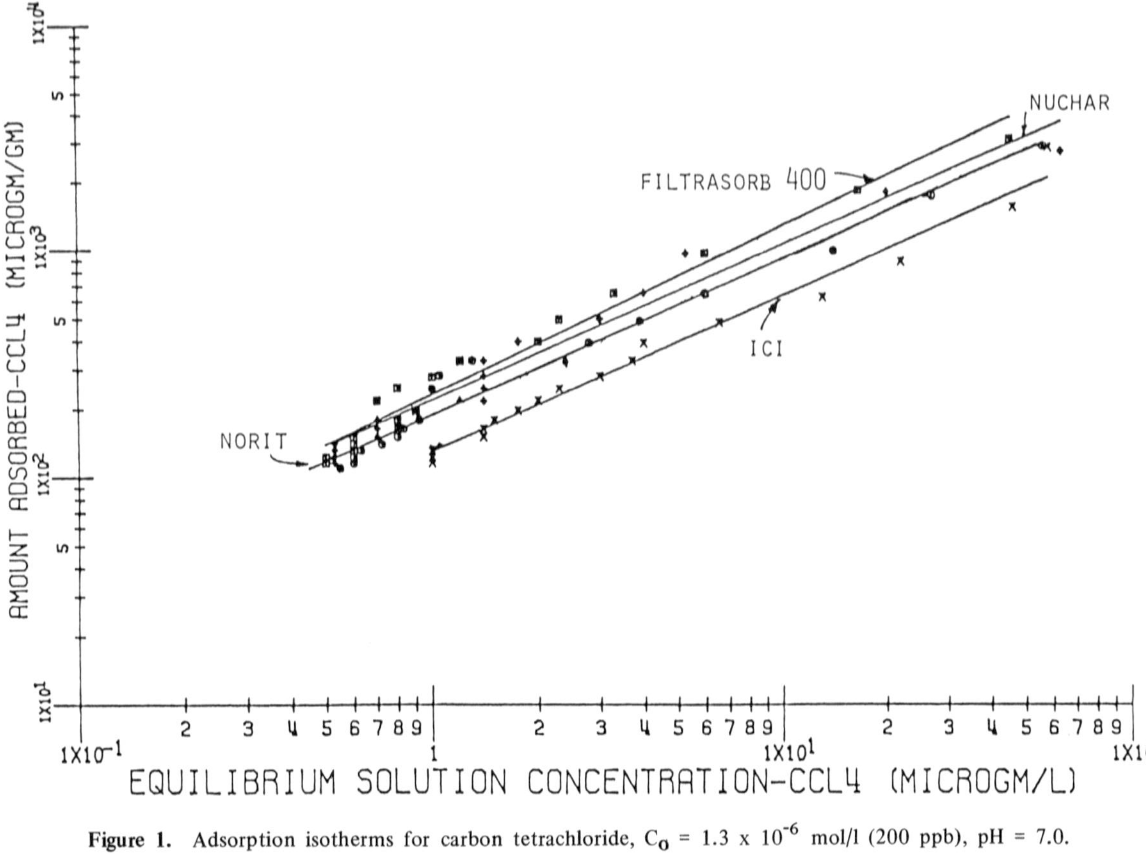

Figure 1. Adsorption isotherms for carbon tetrachloride, $C_o = 1.3 \times 10^{-6}$ mol/l (200 ppb), pH = 7.0.

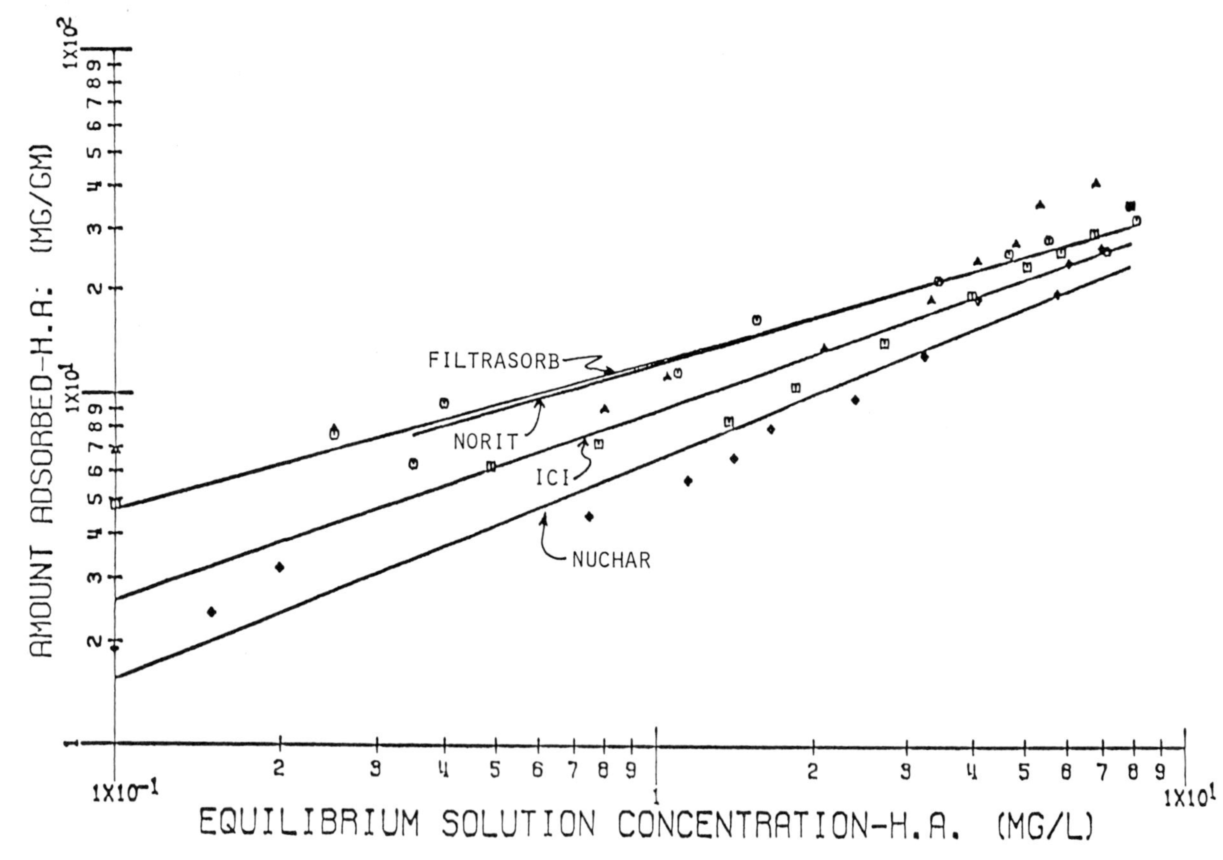

Figure 2. Adsorption isotherms for humic acid, C_o = 8.95 ppm, pH = 7.0.

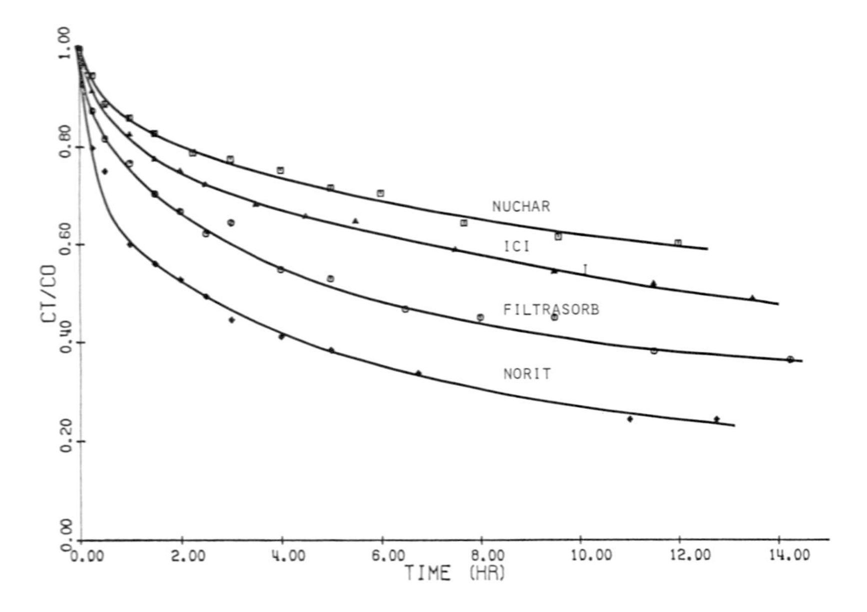

Figure 3. Rates of adsorption of humic acid on different commercial carbons in CMB reactors, C_o = 5 ppm, 16/20 US sieve size carbon, pH = 7.0.

cap, and the vials were agitated by mechanical shaker, at 20°C, until adsorption equilibrium obtained. For each experiment a series of standard solutions containing no carbon was run under identical conditions.

Taking advantage of the volatility of the compounds of interest, a head-space procedure was employed in which the concentration of the liquid phase was quantified by measuring the equilibrium concentration in the vapor phase above it. Pursuant to Henry's Law for dilute solutions, the concentration of a gas dissolved in a liquid at equilibrium is proportional to the partial pressure of that gas in the vapor phase.

Vapor phase concentrations were determined by injecting 0.1 to 0.5 ml of the overhead gas from each vial into the gas chromatograph and comparing detector response with a calibration curve prepared using known weights of compound dissolved in hexane solvent. The amount of compound in the vapor phase of each system containing carbon was then compared with the vapor phase concentrations of the standards to yield, after correction for the vapor-solution equilibrium of the standards, the final aqueous concentration of the sample.

The amount of compound adsorbed on the carbon was determined by subtracting the sum of the amount in the liquid phase and the amount in the vapor phase from the initial amount added to the vial.

Rate Studies

Completely mixed batch (CMB) reactor rate experiments were performed with four selected halogenated hydrocarbons: chloroform, bromodichloromethane, carbon tetrachloride, and bromoform. The experiments were conducted in carefully sealed 2.5-liter glass reactors, a schematic diagram of which is illustrated in Figure 4. A weighed quantity of sieved and washed activated carbon, 16/20 U.S. Standard mesh size, together with the experimental solution, were added to the vapor-phase-free reactor. The carbon was dispersed by a motor-driven Teflon-coated stirrer and 5-ml samples were withdrawn at fixed time intervals. A positive displacement plunger eliminated introduction of any head space by sample volume

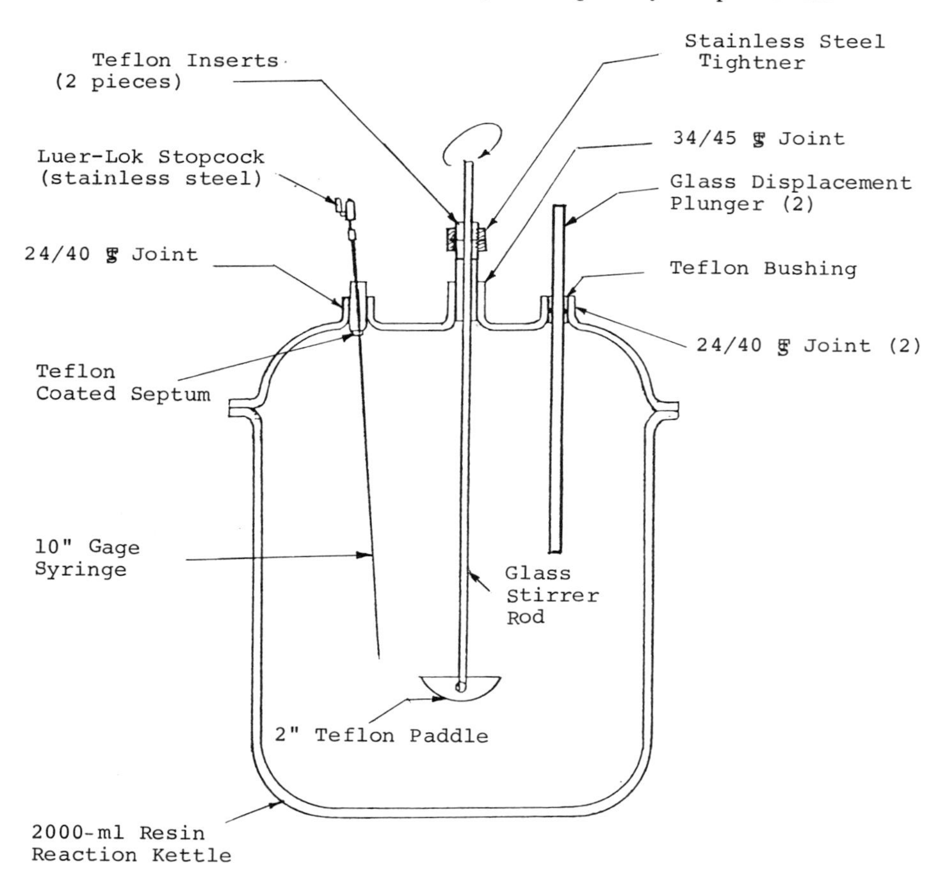

Figure 4. Schematic CMB reactor used for rate studies.

displacement. The samples, together with 2 ml of high-purity hexane, and 1 g of sodium chloride, were placed in 10-ml vials fitted with Teflon-coated screw-on caps. The samples, as well as appropriate standards, were shaken vigorously for 1 min, the hexane allowed to separate and solute analysis performed by injecting a few microliters of the extract into a gas chromatograph equipped with a scandium-tritide electron capture detector. A calibration curve of concentration of the standard solution versus detector response was used for the sample concentration determination.

RESULTS AND DISCUSSION

For mathematical description and quantitation of the adsorption equilibrium data several theoretical and empirical equations, including the Freundlich, Langmuir and a three-parameter isotherm, were investigated. The Freundlich and the three-parameter isotherm were found to provide the best description of the experimental data, and the Freundlich equation was chosen because of its relative simplicity and common use.

The Freundlich equation has the form:

$$q_e = K_F C_e^{1/n} \qquad (1)$$

where q_e = the amount adsorbed per unit weight of adsorbent,
C_e = the amount of solute remaining in solution at equilibrium,
K_F and $1/n$ = characteristic constants relating to adsorption capacity and intensity, respectively.

To quantify the adsorption isotherm parameters, data which accord with Equation 1 are normalized by plotting the logarithm of q_e vs the logarithm of C_e. The data is then statistically (*e.g.*, least squares) fit with a straight line, with slope $1/n$ and intercept $\log K_F$.

Typical adsorption isotherms for chloroform, bromoform, dichloromethane and tetrachloroethene are illustrated in Figures 5 through 8, respectively. The solid lines in these figures represent the best statistical fit of the data by the Freundlich isotherm equation. A tabulation of the Freundlich adsorption parameters for the compounds studied is presented in Table I.

A comparison of the adsorption isotherm constants in Table I with those in Table II for several compounds for which activated carbon is commonly used as a removal technique indicated that activated carbon adsorption capacities and intensities for low-molecular-weight volatile halogenated hydrocarbons at the concentration levels of interest are significantly different from those of higher-molecular-weight compounds.

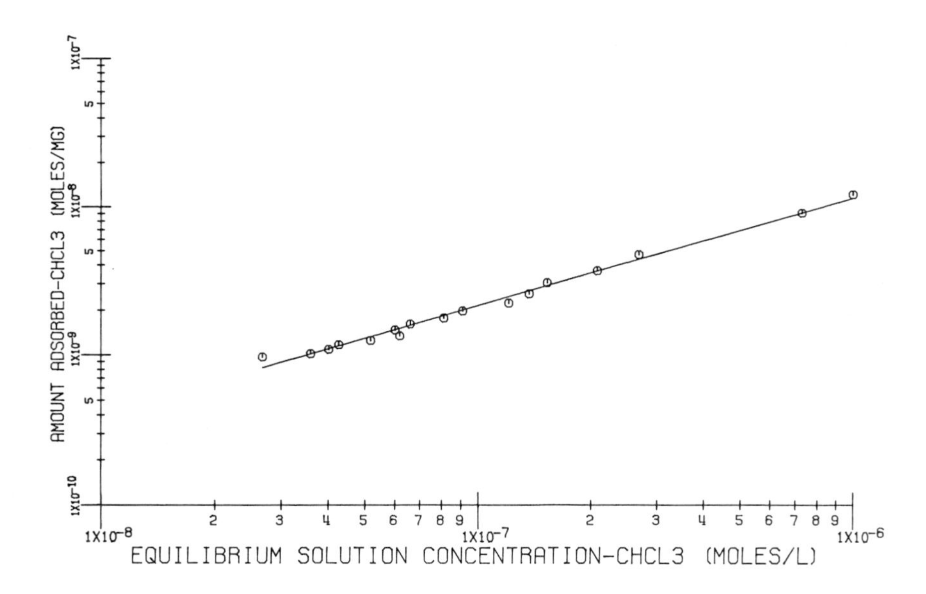

Figure 5. Freundlich isotherm fit of experimental data for chloroform adsorption on Filtrasorb 400 activated carbon, $C_o = 1.67 \times 10^{-6}$ mol/l (200 ppb), pH = 7.0.

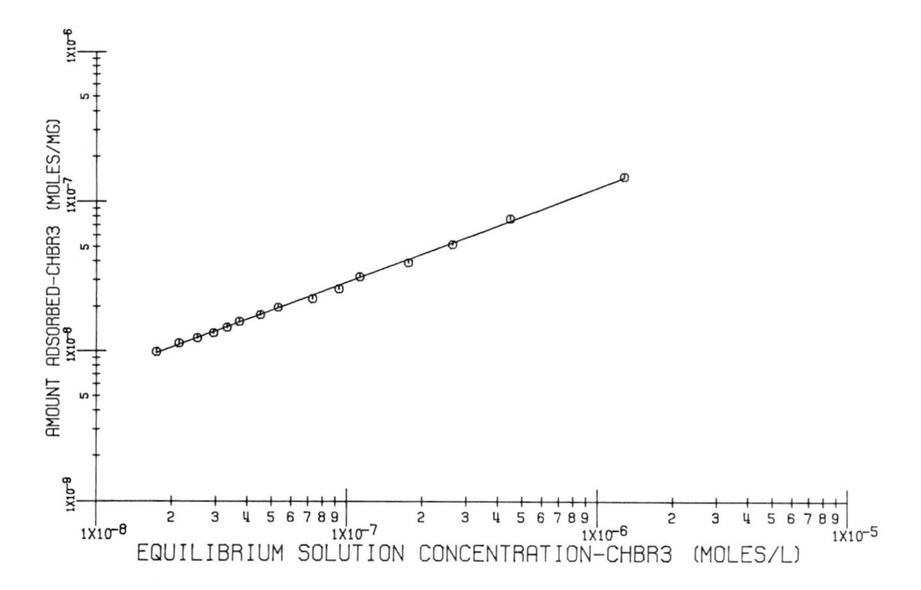

Figure 6. Freundlich isotherm fit of experimental data for bromoform adsorption on Filtrasorb 400 activated carbon, $C_o = 1.58 \times 10^{-5}$ mol/l (4 ppm), pH = 7.0.

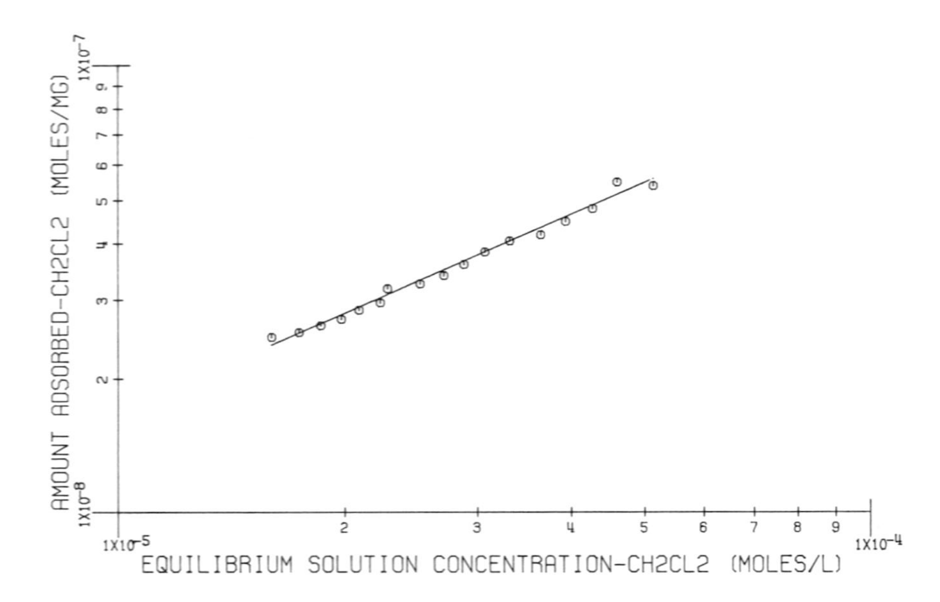

Figure 7. Freundlich isotherm fit of experimental data for dichloromethane adsorption on Filtrasorb 400 activated carbon, $C_o = 5.88 \times 10^{-5}$ mol/l (5 ppm), pH = 7.0.

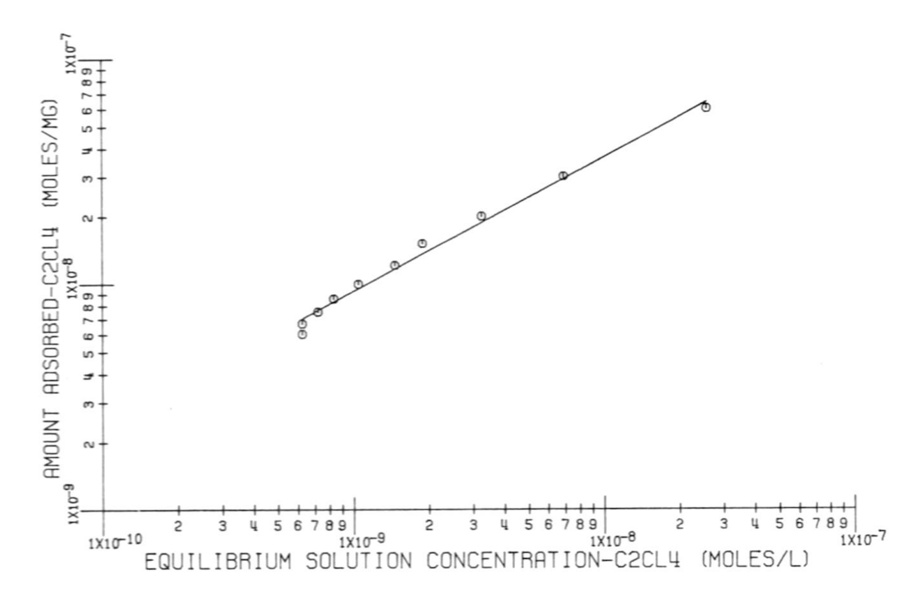

Figure 8. Freundlich isotherm fit of experimental data for tetrachloroethene adsorption on Filtrasorb 400 activated carbon, $C_o = 6.02 \times 10^{-6}$ mol/l (1 ppm), pH= 7.0.

Table I. Freundlich Adsorption Isotherm Constants for Selected Volatile Halogenated Hydrocarbons[a]

Compound	$K_F(\times 10^4)$	1/n	Initial Conc. mol/l$\times10^6$ (ppm)
Chloroform	2.54	0.725	1.67 [0.2]
Bromodichloromethane	11.5	0.745	2.44 [0.4]
Bromoform	6.97	0.626	1.58 [4]
Carbon Tetrachloride	1.6	0.576	1.3 [0.2]
Tetrachloroethene	21.0	0.595	6.02 [1]
Dichloromethane	0.776	0.732	58.8 [5]
Bromochloromethane	0.255	0.574	38.6 [5]

[a]Calculated for q_e in moles/mg and C_e in moles/liter

These differences are partly due to the adsorption characteristics of the individual sorbates and partly due to significant differences in initial concentration of sorbate; the data presented in Table II were measured for initial concentrations of concern for waste treatment applications, or several orders of magnitude higher initial concentration than used to obtain the data presented in Table I.

Table II. Freundlich Adsorption Equilibrium Constants for Selected Organic Compounds of Known Adsorption Characteristics[a]

Compound	$K_F(\times 10^4)$	1/n	Initial Conc. mol/l$\times10^2$ (ppm)
Phenol	0.0461	0.1413	3.0 [2820]
p-Bromophenol	0.0988	0.2006	3.0 [5190]
p-Toluene Sulfonate	0.0129	0.1474	3.0 [5130]
Dodecyl Benzene Sulfonate	0.043	0.1812	3.0 [9750]

[a]Calculated for q_e in moles/mg and C_e in moles/liter.

The results of the rate studies are presented in Figure 9. The procedure devised and performed for the rate studies with these volatile compounds was particularly successful, partly because of the elimination of solute loss due to evaporation. Furthermore, examination showed that since the solution in the CMB reactor was in contact only with glass and Teflon surfaces, no measurable sorbate loss to the containing surfaces was encountered. Elimination of these losses is extremely important when rate studies are performed with very dilute solutions of volatile compounds.

Figure 10 presents typical rate data for organic compounds commonly removed from wastewaters by activated carbon (Crittenden and Weber, 1976). Again it is clear that the chemical character of these types of

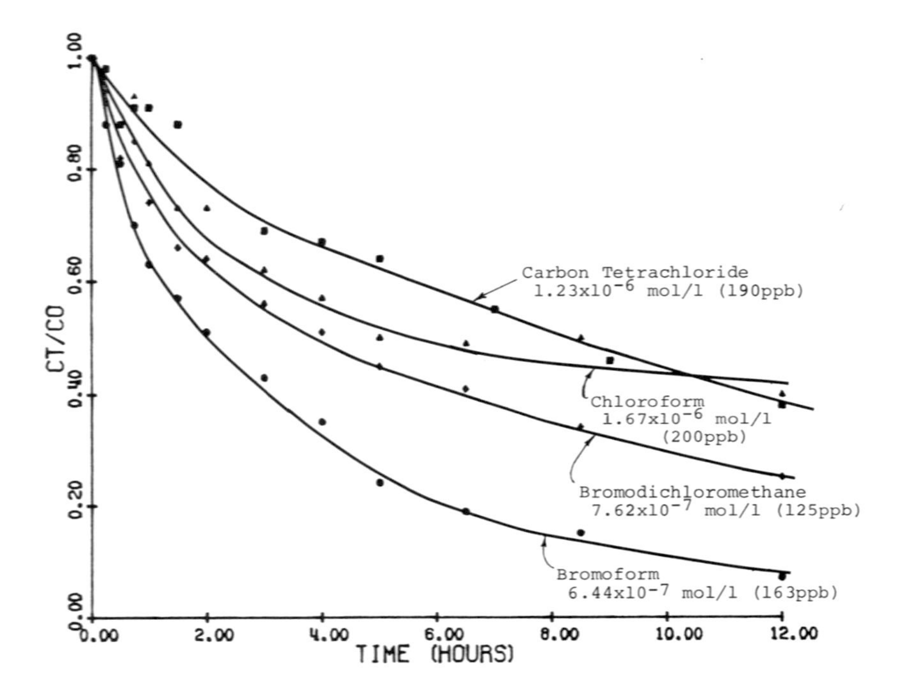

Figure 9. Rates of adsorption of typical volatile halogenated compounds in CMB reactors, 16/20 US sieve size Filtrasorb 400 activated carbon; carbon = 0.19 g/l, pH = 7.0

substances, coupled with their higher concentrations, lead to different adsorption characteristics than those of the volatile halogenated hydrocarbons of concern for water treatment.

The essential conclusion then is that, while adsorption on activated carbon can provide removal of low-molecular-weight volatile halogenated compounds, it is necessary that appropriate fundamental information regarding the specific adsorption characteristics of such substances–under appropriate conditions–be developed to facilitate rational design and operation of adsorption facilities for water treatment applications. This is the objective of the ongoing research program at the University of Michigan.

ACKNOWLEDGMENT

The research reported herein has been sponsored in part by Research Grant Number R-804269 from the Municipal Environmental Research Laboratory, U.S. Environmental Protection Agency. The authors express

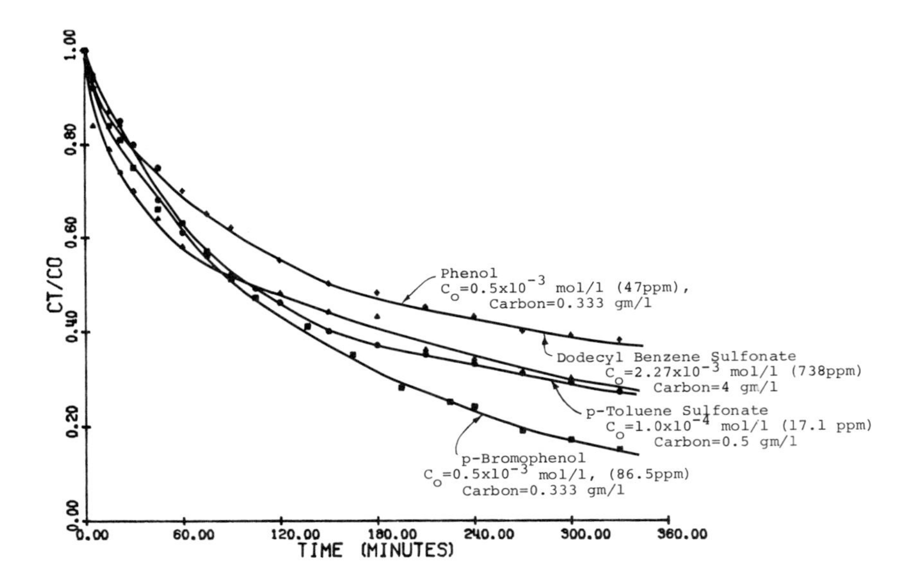

Figure 10. Rates of adsorption of typical wastewater compounds in CMB reactors, 16/20 US sieve size Filtrasorb 400 activated carbon, pH = 7.0.

their appreciation to Ms. Gail Melson for her able technical assistance in the laboratory work described herein.

REFERENCES

Bellar, T. A. and J. J. Lichtenberg. "The Determination of Volatile Organic Compounds at the μg/l Level in Water by Gas Chromatography," USEPA, Cincinnati, Ohio (November 1974).

Coleman, W. E., R. D. Lingg, R. G. Melton and F. C. Kopfler. "The Occurrence of Volatile Organics in Five Drinking Water Supplies Using Gas Chromatography/Mass Spectrometer (GC/MS)," *USEPA Report* (June 1975).

Crittenden, J. C. and W. J. Weber, Jr. "A Predictive Model for Design of Fixed-Bed Adsorbers: I. Model Development and Parameter Estimation; II. Single-Component Model Verfification; III. Multicomponent Model Verification," presented at Am. Soc. Civ. Engr. Specialty Conference, Seattle, Washington, 1976, *J. Env. Engr. Div., Proc. ASCE* (In Press).

Dowty, B., D. Carlisle, J. L. Laseter and J. Storer. "Halogenated Hydrocarbons in New Orleans Drinking Water and Blood Plasma," *Science* 187:75 (1975).

Heuper, W. C. and W. W. Payne. "Carcinogenic Effects of Adsorbates of Raw and Finished Water Supplies," *Am. S. Clin. Path.* 39(5):475-481 (1963).

Mieure, J. P. "A Rapid and Sensitive Method for Determining Volatile Organohalides in Water," *J. Am. Water Works Assoc.* (January 1977).
Morris, R. L. and L. G. Johnson. "Agricultural Runoff as a Source of Halomethanes in Drinking Water," *J. Am. Water Works Assoc.* (September 1976).
Nicholson, A. A. and O. Meresz. "Organics in Ontario Drinking Water: Part I. The Occurrence and Determination of Free and Total Potential Haloforms," Ontario Ministry of the Envir. Lab. Service Branch, presented at Pittsburgh Conf. of Anal. Chem. and Applied Spectroscopy (1976).
N.C.I. Report on the Carcinogenesis Bioassay of Chloroform, Carcinogenesis Program, Division of Cancer Cause and Prevention (Bethesda, Maryland: National Cancer Institute, March 1, 1976).
Ongerth, H. J., P. Spath, J. Crook and E. Greenberg. "Public Health Aspects of Organics in Water," *J. Am. Water Works Assoc.* (July 1973).
Robeck, G. G. "Evaluation of Activated Carbon," A Report to the Water Research Laboratory, "Organics in Drinking Water," USEPA, Cincinnati, Ohio (March 3, 1975).
Rook, S. S. "Formation of Haloforms During Chlorination of Natural Waters," *Water Treatment and Examination* 23(2):234 (1974).
Rook, S. S. "Haloforms in Drinking Water," *J. Am. Water Works Assoc.* (March 1976).
Stevens, A. A., C. S. Slocum, D. R. Seeger and G. G. Robeck. "Chlorination of Organics in Drinking Water," USEPA Report, Cincinnati, Ohio (1975).
Tardiff, R. G. and M. Deinger. "Toxicity of Organic Compounds in Drinking Water," *Proc. of 15th Annual Water Quality Conf.* The University of Illinois (1973).
USEPA. "Draft Analytical Report–New Orleans Area Water Supply," Dallas, Texas (November 1974).
USEPA. "National Organic Reconnaissance Survey for Halogenated Organics in Drinking Water," Cincinnati, Ohio (April 15, 1975).
Weber, W. J., Jr. and J. C. Morris. "Equilibria and Capacities for Adsorption on Carbon," *J. San. Eng. Div., Proc. ASCE* 90:SA3 (1964).
Weber, W. J., Jr. *Physicochemical Processes for Water Quality Control,* (New York: Wiley Interscience, 1972).
Weber, W. J., Jr. "Purification of Industrial Effluents with Activated Carbon," *Proceedings, NATO Advanced Study Institute: Scientific Aspects of Sorption and Filtration Methods for Water Purification,* Institut for Atomenergi, Kjeller, Norway (1974a).
Weber, W. J., Jr. "The Prediction of the Performance of Activated Carbon for Water Treatment," *Activated Carbon in Water Treatment,* The Water Research Association, Medmenham, Marlow, Bucks, SL72HD, England (February 1974b).
Weber, W. J., Jr. and W. Ying. "Integrated Biological and Physicochemical Treatment for Reclamation of Wastewater," paper presented at International Conference on Advanced Treatment and Reclamation of Wastewater, Johannesburg, South Africa (June 1977).

SECTION III

INORGANICS

9

HEAVY METALS–TOXICITY TO AQUATIC BIOLOGICAL FORMS AND TO MAN

Rolf Hartung, Ph.D.
Professor of Environmental Toxicology
Department of Environmental & Industrial Health
The University of Michigan
Ann Arbor, Michigan

INTRODUCTION

Toxicology may be defined as the study of the harmful effects of chemicals on biological organisms, including the evaluation of their safety. The concept of safety is a particularly difficult one to treat. I am convinced that the toxicologist engaged in safety evaluations, the legal profession and consumers in general have a similar impression of the concept of safety. However, when this concept is examined in detail, significant differences become apparent. The legal profession and the public frequently appear to view the concept of safety as an absolute guarantee that nothing adverse will happen when any person or organism is exposed to a specific chemical substance. However, from the toxicologist's viewpoint, an equivalent guarantee of safety would mean that he would have to prove a conceivably possible event will not happen at a specific dose level. Regrettably, this means the toxicologist must contend with a logical paradox. It is neither possible nor desirable to expose all organisms, including man, to all possible combinations and dose levels of a toxicant to establish at what dose levels and in which manner they will react. Therefore, the toxicologist is forced to make estimates on the basis of incomplete experimental data, which may have been gathered in species other than man. As a result, he may be able to give reasonable assurances of safety, but be unable to guarantee it.

Another important concept in toxicology is that the effects of various substances cannot be considered without reference to dose. However, in common usage we view some substances as poisonous and others as harmless, without reference to dose. I have often tested the extent to which this thinking is engrained in our society by asking the following questions during lectures to relatively large groups of people:

1. How many of you consider arsenic a toxic substance?
2. How many of you consider sugar a toxic substance?
3. How many of you have personally been affected by arsenic poisoning or have had close relatives or close friends who were affected by arsenic?
4. How many of you have consumed personally, or have had a close relative or close friend who has consumed, sufficient hard candy or sugar to have become nauseated, dizzy and have vomited; or, in other words, have experienced the symptoms of acute toxicity due to sugar?

The outcome of this highly simplified line of questioning is predictably the same. Almost everyone considers arsenic a toxic substance and sugar to be harmless. However, almost nobody knows of anyone who has been subject to arsenic poisoning while almost everyone knows someone who has been affected by sugar poisoning. While this line of questioning does little to explore the toxicology of sugar in relation to arsenic, it clearly illustrates that nearly all of us think of toxicity without relating it to dose. This attitude is prevalent throughout our society including scientists, lawyers, legislators and consumers. As a consequence, scientific discussions of toxicity and evaluations of safety often become clouded by emotional issues.

However, as well as can be determined, there is a sufficiently large amount of any substance at which every individual of every species would show an effect. Likewise, there is a dose sufficiently small where no effects can be measured, no matter how sophisticated the measurement, no matter how many individuals are being tested. It should be noted, however, that a "no measured effect" level is not necessarily a safe level. The "no measured effect" level addresses itself merely to the organisms that were actually exposed to the chemical; it does not predict outside the range of experimental results. The tests can only allow guesstimates but cannot predict whether a much larger population, such as the entire population of the United States, would be expected to show no effect. It also provides no more than a scientifically educated guess as to how another species might react.

Even though dose is important, it is not always easily determined. A number of factors influence dosage, such as the concentration of the toxicant, duration of exposure, chemical form, particle size, pH, alkalinity, hardness and other factors (National Academy of Sciences, 1974). When discussing heavy metals, the tendency has been to ignore especially the chemical form of the metal. However, the toxicities of the various forms may differ significantly.

For example, the oral LD_{50} of mercurous chloride in rats is 210 mg/kg of body weight, while that of mercuric chloride is 37 mg/kg, and that of methylmercuric dicyandiamide is 32 mg/kg (NIOSH, 1974). On chronic exposure, the relative toxicity of the organic mercury compound increases significantly (Hartung and Dinman, 1972).

A metal such as copper does not exist just as copper but in a multitude of chemical forms, including ionic copper. The toxicity of every one of these forms is different (Andrew, 1976).

Many toxicological studies are done with one particular compound in mind so that the researcher may be dosing an aquarium with copper nitrate with the expectation that the exposure of aquatic organisms would be due to copper ions. However, within fractions of a second after application of the test substance to the test tank, the ionic copper would be changed into a multitude of hydroxides, carbonates and other more complex forms, including a small residual of ionic copper. Thus, the observed effects are not due solely to ionic copper but to a complex mixture of copper salts, which may or may not duplicate environmental conditons.

Some of the effects, such as lethality, produced by toxic substances are obvious. However, lower doses of toxicants may produce effects that can also affect the long-term survival of a species, and the effects under these conditions may not be as readily studied as lethality. Such chronic effects may include those on growth rate, life span, behavioral changes, reproduction and changes in the ecological community structure.

In addition, a number of factors influence the responses or effects seen. Among these are the actual dose absorbed, the body burden, potentiation and antagonism with other toxicants present, the degree of acclimatization to the toxicant, the previous health experience of the organism, the sex and age of the organism and, most importantly, the species involved.

The question of species is particularly important. We are all aware of the complexities involved in trying to establish the safety of a toxicant to man through studies in laboratory animals. The selection of representative species for environmental safety evaluations is much more complex. The number of species that comprise an ecological community are obviously quite numerous. It is assumed that many of these have significant ecological function. However, with present constraints of manpower and finances, it is not possible to even test representatives of all the major orders.

The heavy metals are thought to act through several types of mechanisms. For acute effects, the most important mechanisms in fish appear to be due to coagulation of mucous membranes, especially in the gills. Many of the heavy metals are potent inhibitors of enzymes, especially through bonding with sulfhydryl groups, which frequently are at the reactive centers of many enzymes. This mechanism is particularly important for silver, mercury, lead and cadmium. In addition, the heavy metals can produce relatively insoluble soaps by combining with fatty acids. The significance of this mechanism is not well understood in relation to the more predominant mechanism of sulfhydryl bonding. A number of special mechanisms, such as the inhibition of heme synthesis by lead, have also been identified. But overall, the number of mechanisms by which heavy metals can act appears to be relatively limited. Nevertheless, the toxicity of the various heavy metals may differ significantly as illustrated in Table I, which summarizes current water quality criteria. In a number of instances, these water quality criteria recognize that many factors influence the chronic toxicity of the heavy metals so that the criteria are expressed as fractions of the 96-hr/LC_{50} value in a sensitive species, determined in the receiving water. Table I indicates that in many instances the water quality criteria proposed by the International Joint Commission (1976) differ from those proposed by the Environmental Protection Agency (1976).

Table I. Proposed Water Quality Criteria

	EPA (mg/l)	IJC (μg/l)
Cd	0.4 soft water–salmonids	0.2
	1.2 Hard water–salmonids	
	4.0 soft water–other	
	12.0 hard water–other	
Cr	50 (D)[a]	50 (D)[a]
	100 (F)	
Cu	1000 (D)[a]	5
	0.1 x 96 hr LC_{50}	
Pb	50 (D)[a]	10–Lake Superior
	0.01 x 96 hr LC_{50}	20–Lake Huron
		25–other Great Lakes
Hg	2.0 (D)[a]	0.2 and 0.5 (mg/g) fish concentration
	0.05	
	0.10 (M)[b]	
Ni	0.01 x 96 hr LC_{50}	25
Ag	50 (D)[a]	
	0.01 x 96 hr LC_{50}	
Zn	5000 (D)[a]	30
	0.01 x 96 hr LC_{50}	

[a](D) = domestic water.
[b](M) = marine biota.

In both instances, the agencies are looking at the same basic scientific data to derive their proposed criteria. Why, then, the differences? It must be recognized that there is latitude for interpretation of the scientific data. In some instances it is difficult to ascertain whether a particular effect may be harmful. In addition, there may be differences in the philosophy as to how stringently particular findings may be applied. Also, the water quality objectives proposed by the International Joint Commission are designed to apply over a much narrower set of environmental conditions, namely the Great Lakes.

The differences also reflect the difficulties in developing water quality criteria designed to apply to a wide range of environmental conditions. With the present state-of-the-art, it is exceedingly difficult to develop water quality criteria that can account for all the variables, and actually it may not be possible to express them as single numbers.

It is impossible here to discuss in detail the toxicology of all the heavy metals to aquatic organisms and to man; therefore, I will concentrate on one heavy metal, mercury, to provide at least some indication of the complexity of the information that must be considered to establish a criterion that can provide a reasonable assurance of safety. A compound of major interest in the toxicology of mercury is methylmercury, which may be formed by at least two separate mechanisms from ionic mercury (Wood *et al.*, 1968; Landner, 1971).

Methylmercury may also be degraded into metallic mercury (Spangler *et al.*, 1973) so that the amount of methylmercury present in the environment at any one time is due to a critical balance between methylation and demethylation processes. The environmental chemistry of mercury has been greatly simplified in Figure 1. This facet alone could be the subject of a major monograph.

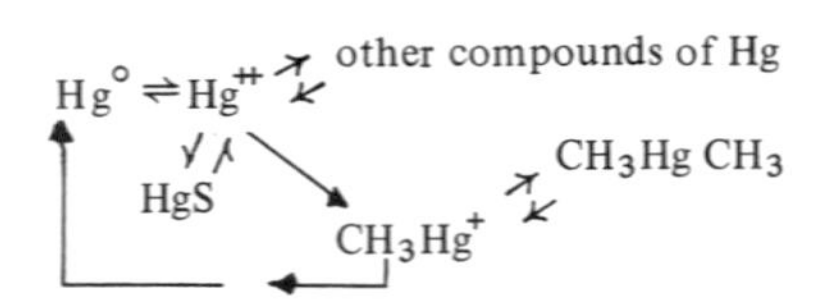

Figure 1. Simplified overview of known environmental transformations of mercury.

Several instances of poisoning by methylmercury in man have been well documented. These incidences can be divided into two major types: those in which man has ingested seed grain that has been treated with a methylmercury compound as a fungicide, and those in which man has consumed large amounts of seafood that have concentrated excessive quantities of methylmercury. I have listed in Table II the symptoms of alkylmercury poisoning according to reversibility and escalating severity. The potential severity and irreversibility of some of these symptoms illustrates some of the reasons why methylmercury has generated so much concern as a potentially hazardous compound. Examination of Table II should make it clear that early symptoms may go unrecognized and may not be particularly helpful in epidemiological studies, since many different causes may produce similar results.

Table II. Progression of Symptoms of Alkyl Mercury Poisoning in Man[a]

Symptoms
Reversible Symptoms
Headache
Fatigue
Inability to concentrate
Impairment of memory
Possibly Reversible Symptoms
Blurred vision
Tingling of fingers
Numbness of fingers
Impairment of fine finger coordination
Impairment of sensation in the corners of the lips
Emotional irritability
Irreversible Symptoms
Significant loss of finger coordination
Loss of motor control:
hands, locomotion, speech
Sensory defects:
auditory, visual field, patchy areas of impairment of sensation to touch, blindness

[a]Adapted in part from Dinman and Hecker (1972).

While it is likely that the proportion of the total amount of mercury in the environment that is in the form of methylmercury at any one time is minute, due to the fact that it exists are a result of a balance between methylating and demethylating systems, this compound nevertheless is of primary concern to the water resources scientist and toxicologist. The major reason for this is that freshwater fish are able to absorb methylmercury preferentially, so that nearly their entire body burden of total mercury occurs in the form of methylmercury. Thus, the major source of exposure of people to methylmercury arises from the consumption of fish. Similarly, the toxicity of methylmercury to aquatic organisms and to fish-consuming mammals and birds can be considerable (Table III). The exposures of man in North America have been regulated by the U.S. Food and

Table III. Mercury Toxicity Studies[a]

Organism	Compound	Effect	Lowest Concentration Producing Effect	Highest No-Effect Concentration	Remarks
Brook Trout	CH_3Hg^+	Cough response	3 μg/l		5-day exposure
Rainbow Trout	Hg^{++}	Decreased activity	50 μg/l		4-6-day exposure
Brook Trout	CH_3Hg^+	Deformities and deaths in 2nd generation	0.93 μg/l	0.29 μg/l	3-generation exposure
Cat	CH_3Hg^+	CNS deaths	0.25 mg/kg/day		55-96-day feeding
Mallard	CH_3Hg^+	decreased hatching	3 μg/g (diet)	0.5 μg/g (diet)	21-week exposure
Mallard (duckling)	CH_3Hg^+	increased avoidance response	0.5 μg/g (diet)		Ducks fed prior to and during reproduction phase

[a]Excerpted from International Joint Commission Report (1976).

Drug Administration and Canadian Food and Drug Directorate by setting action guideline levels of 0.5 ppm of total mercury in edible portions of fish. Efforts are underway to ensure the quality of the aquatic environment by the application of various water quality objectives or water quality criteria. These are based on studies of the type briefly summarized in Table III, which has been excerpted from the International Joint Commission Water Quality Objectives for the Great Lakes (1976). It should be noted that very low quantities of methylmercury can produce effects in fish during long-term exposures. However, it is not easy to select a water quality criterion for total mercury in water that will protect all beneficial uses of water, primarily because it is not known what proportion of the total mercury in water is indeed in the form of methylmercury. While an exposure of fish to 0.2 μg/l of methylmercury is probably safe for many species of fish, these fish will still bioconcentrate methylmercury to levels far in excess of 0.5 ppm during such an exposure. Such high concentrations of methylmercury in fish are clearly not acceptable for consumption by humans or fish-eating mammals and birds.

Since it is not possible to determine the concentration of methylmercury in water in environmental samples, it is not possible to predict the amount of bioaccumulation of methylmercury in aquatic organisms from measurements of total mercury concentrations in water alone. However, the level of bioaccumulation in aquatic organisms is vital to assuring the safety of aquatic organisms as a source of food for man and wildlife. As a consequence, the IJC proposed a combined standard that would ensure short- to medium-term protection of the aquatic organisms by setting a water quality objective of 0.2 μg/l total mercury in finished water based on a no measured effect level determined in a three-generation study in brook trout exposed to methylmercury. An additional requirement set by the IJC was that the concentration of bioaccumulated mercury in whole fish should not exceed 0.5 ppm on a wet weight basis, in order to protect fish-eating mammals and birds. Obviously, both requirements must be fulfilled to meet the total water quality objective proposed by the IJC for mercury.

The approach taken by the Environmental Protection Agency to the evaluation of the environmental toxicology and safety evaluation of mercury is somewhat different from that taken by the International Joint Commission. The EPA is faced with the more difficult problem of developing water quality criteria that are applicable throughout the United States, rather than being applicable only to a local region. In addition, the EPA has sought to establish water quality criteria that would protect all uses, including those which would be compromised by bioaccumulation, by setting single water quality criteria. In the specific case of mercury, the EPA document on Quality Criteria for Water (1976) also considers the preëminent toxicity and environmental behavior of methylmercury. The EPA then derives its criterion by considering an experimentally derived bioaccumulation factor of 10,000 for methylmercury in conjunction with the FDA action guideline of 0.5 μg/kg in fish. Assuming that all mercury in water is in the form of methylmercury, and assuming that mercury absorbed onto particulates suspended in water is as readily available to fish as dissolved forms of mercury, the EPA arrives at a water quality criterion for total mercury in surface water of 0.05 μg/l to ensure the protection of freshwater aquatic life and wildlife.

In conclusion, the study of the environmental toxicology of heavy metals with reference to effects and safety evaluations for aquatic organisms and man, is exceedingly complex. This complexity is caused by subsidiary problems in the definition of safety, the intricacies of the environmental chemistry and environmental dynamics of all heavy metals, as well as the large number of species that must be tested for a multiplicity of complicated toxicological effects. As a consequence, our present efforts in environmental toxicology are often inadequate to produce sufficient relevant data to permit the development of environmental quality criteria that are fully defensible on a scientific basis.

REFERENCES

Andrew, R. W. "Toxicity Relationships to Copper Forms in Natural Waters," in *Toxicity to Biota of Metal Forms in Natural Water,* R. W. Andrew and P. V. Hodson, Eds. (Windsor, Ontario: International Joint Commission, 1976), pp. 127-144.

Dinman, B. D. and L. H. Hecker. "The Dose-Response Relationship Resulting from Exposure to Alkyl Mercury Compounds," in *Environmental Mercury Contamination,* R. Hartung and B. D. Dinman, Eds. (Ann Arbor, Michigan: Ann Arbor Science Publishers, Inc., 1972), pp. 290-301.

Hartung, R. and B. D. Dinman, Eds. *Environmental Mercury Contamination* (Ann Arbor, Michigan: Ann Arbor Science Publishers, Inc., 1972), pp. xi, 349.

International Joint Commission (IJC). *Proposed New and Revised Water Quality Objectives* (Windsor, Ontario: International Joint Commission, 1976), pp. iv, 182.

Landner, L. "Biochemical Model for the Biological Methylation of Mercury Suggested From Methylation Studies *in Vivo* with *Neurospora crassa,"* *Nature* 230:452-453 (1971).

National Academy of Sciences (NAS). *Water Quality Criteria* (Washington, D.C.: U.S. Government Printing Office, 1974), pp. xix, 594.

National Institute for Occupational Safety and Health (NIOSH). *The Toxic Substances List* (Rockville, Maryland: National Institute for Occupational Safety and Health, 1974), pp. xciv, 904.

Spangler, W. J., J. L. Spigarelli, J. M. Rose and H. M. Miller. "Methylmercury: Bacterial Degradation in Lake Sediment," *Science* 180:192-193 (1973).

United States Environmental Protection Agency. "Quality Criteria for Water," EPA Publication 440/9-76-023 (1976), pp. ix, 501.

Wood, J. M., F. S. Kennedy and C. G. Rosen. "Synthesis of Methylmercury Compounds by Extracts of a Methanogenic Bacterium," *Nature* 220:173-174 (1968).

10

THE USE OF ATOMIC ABSORPTION IN WATER QUALITY ANALYSIS

Richard D. Ediger
Perkin-Elmer Corporation
Lombard, Illinois

The analytical problems for the water chemist have been increasing over the past few years. Pressure is being put on the analyst to provide more accurate numbers for more water components at lower concentrations. It seems that every several months he is asked to add a new analysis to his list or to measure a presently determined component at a previously unheard-of detection limit. With the continuing emphasis on the health effects of toxic substances in our potable water supplies, this trend will no doubt continue. Within recent years, we have had to deal with mercury, cadmium and, now selenium. Elements never considered in the past now have a high priority on the water chemist's list of water quality parameters.

Fortunately, though, the analyst does not have to throw his hands up in disgust at these new requirements–for as the requirements have been increasing, so have the techniques for meeting these requirements. Taking the area of metals analysis, for example, procedures utilizing atomic absorption spectrophotometry have made the routine determination of such elements as calcium, sodium and potassium much easier than they have ever been before. In addition, the same technique has made possible the determination of such trace components as lead, copper and mercury at levels much below 1 mg/l. Atomic absorption has been used quite routinely in such analytical disciplines as agriculture, metallurgy and medicine for 10 to 15 years. Its use in water quality laboratories has taken a sudden surge forward during the past five years. It is used not only in research laboratories of the state and federal regulatory agencies, but also in laboratories of small water districts for monitoring the quality of potable water supplies and the effluent from municipal water and sewage treatment facilities.

Atomic absorption is now considered the method of choice for the determination of almost all metallic elements in water. Its analytical range covers major water components such as calcium and sodium at several hundred milligrams per liter to the ultratrace components such as mercury at less than 0.0001 mg/l. It is directly applicable to about 65 different elements as well as to a few components such as sulfate, phosphate and chloride by indirect procedures. In most cases little or no sample preparation is necessary and the volume of sample used is generally a few milliliters or less. The use of the instrument is straightforward and does not necessarily require a degree in chemistry for its proper operation. Most operators with a little water lab experience can be trained to generate useful results in a few hours.

An added attraction is that standardized procedures for the metals of interest in water quality analysis are readily available. Federal agencies such as the Environmental Protection Agency and U.S. Geological Survey have procedure manuals available, and several instrument manufacturers have been actively pursuing the development of new methods (EPA 1974; Brown *et al.,* 1970). In general atomic absorption can be characterized as the easiest method for the determination of metals in water samples at either major or trace levels.

INSTRUMENTAL REQUIREMENTS

Most of us are familiar with the use of a simple colorimeter or spectrophotometer in the water laboratory. With these instruments, we are determining the amount of a substance in a solution by measuring how much light is absorbed by the molecules of the substance. With an atomic absorption spectrophotometer, we are doing a similar measurement, except that we are measuring the absorption of light by atoms of the element being determined instead of by molecules containing the element. Whereas molecules generally absorb light over a broad range of wavelengths, atoms will absorb at only a few very specific wavelengths. The atom must exist as a free atom unbound to any other element and be unionized before it will absorb.

These principles dictate several basic requirements for an atomic absorption instrument. First, a sampling "cell" or device is needed to break apart the atoms from their chemical bonds and to keep them in the ground state for absorption. Most often an air-acetylene flame is used as the "cell." It serves to convert the element dissolved in solution to free atoms that may absorb light. In cases where extreme sensitivity is required, the flame is replaced by more efficient sampling devices.

The next instrumental requirement is a source of the narrow wavelengths of light to be absorbed by the atoms of the element being determined. A hollow cathode lamp is usually used as the source. This type of lamp is filled with neon or argon at low pressure and has a cathode formed of the same element being determined. Passage of current through the lamp causes the region near the cathode to emit spectral lines or wavelengths specific for the cathode material. The atoms of the sample in the flame then absorb light at certain of these wavelengths, called resonance lines, and this absorption is then related to the concentration of the element in the sample. For several elements such as arsenic and selenium, alternate sources called electrodeless discharge lamps offer distinct improvements over the performance of hollow cathode lamps. Since we are interested in absorption at only one of the many wavelengths emitted by the source lamp, the next component of the instrument, the monochromator, isolates that wavelength from all the others and passes the light onto the photomultiplier. This senses the light and converts it to electrons. The photomultiplier sees only the isolated resonance wavelength and any absorption of its light by the sample atoms. After suitable manipulation of the photomultiplier signal by the instrument electronics, the analytical results are displayed to the operator. Generally, the sample concentration is displayed on the instrument's digital readout, but other readout devices such as meters, recorders, printers and even teletypewriters may be used.

Many new atomic absorption instruments incorporate a microcomputer directly into the instrument to make operation simpler and more rapid. Although usable spectrophotometers may be purchased for about $6,000, most laboratories are finding that units in the $8,000 to $12,000 price range more fully suit their analytical requirements.

FLAME ATOMIZATION PROCEDURES

Generally, an air-acetylene flame is used as an atomization cell, serving to transform a sample in solution to the atoms necessary for the absorption measurement. With this system, the sample is aspirated through a nebulizer and is converted into a very fine mist. This mist enters a chamber where it is mixed with the air and acetylene gases. The smallest of the sample droplets pass through a slot in the titanium burner head and burn in a long, narrow flame. The thermal energy provided by the flame serves to evaporate the water from the sample droplet, causes it to disintegrate and then breaks down the chemical compounds into their component atoms, which may absorb light from the lamp. In addition to the air-acetylene flame used for many elements, a higher temperature nitrous oxide-acetylene flame is required for the determination of elements such as aluminum and silicon which form heat-resistant compounds with oxygen in the air-acetylene flame. The nitrous oxide-acetylene flame is also useful in some situations to eliminate certain interferences.

For most routine water samples, operating procedures are simple and the analysis proceeds rapidly. The lamp for the element being determined is installed in the spectrophotometer and the instrument is set to the correct wavelength. The flame is ignited and the analysis may begin. Standard solutions covering the expected concentration range of the samples are prepared. Generally, the standards are used to calibrate the instrument to read out the concentration of the sample directly on the digital display without the necessity of plotting a calibration curve on graph paper.

With modern instruments containing internal microcomputers (generally called microprocessors), calibration is simple. The first step is to tell the instrument the concentration of the standard or standards being used. This is done by entering the concentration values into the instrument with a calculator-type keyboard and pressing the appropriate standard calibration button to store the value into the computer memory. After setting the digital display to zero by aspirating a blank solution and depressing the auto-zero function key, the standards are aspirated and the appropriate calibration buttons are again depressed to enter the absorption values of the standards into the instrument. The microprocessor then automatically constructs an accurate calibration curve and stores it in its memory. All we need to do then is to sequentially aspirate our samples and the instrument will calculate their concentrations and display them to us.

If all our samples are in the normal working range of the instrument and the sample absorbance varies linearly with sample concentration, we only have to enter one standard into the instrument. For higher concentrations, however, the calibration curve may bend and more than one standard may be required. For small curvatures, we need enter only two standards; for large deviations from linearity, three standards may be entered and the microprocessor will automatically calculate the appropriate curve and ensure that our calibration is accurate.

POTENTIAL INTERFERENCES

One of the advantages of atomic absorption techniques is the lack of interferences compared to some other methods of analysis. The interferences that do occur are well understood, and it is relatively easy to control them. Potential interferences are of several types: chemical, matrix, ionization and background absorption.

Chemical interferences occur when the element being determined combines chemically with another reactive component in the sample. The resulting compound influences the atomization process in the flame and thus alters the number of free atoms available to absorb light. There are two methods commonly used to control chemical interferences: the addition of a releasing agent to the samples and the use of a higher temperature flame. A releasing agent is a chemical which reacts preferentially with either the element being determined or with the interfering component. The most common example is the addition of lanthanum chloride to calcium solutions to overcome the effect of phosphate on the calcium signal. The use of a higher temperature flame overcomes many interferences because there is more energy to break down compounds which would be stable in cooler flames. The interference of phosphate in calcium observed in the air-acetylene flame is removed in the nitrous oxide-acetylene flame.

Matrix interferences occur when physical properties such as viscosity or surface tension differ between the sample and standards. These interferences may occur when samples contain a high concentration of dissolved salts or when the sample and standards are prepared with different solvents. This interference is most often controlled by matching the dissolved salt concentration and the solvent in the samples and standards or by diluting the sample until the dissolved salt effect is negligible.

Ionization interferences occur when the flame temperature is high enough to ionize a significant fraction of the element being determined. This reduces the number of atoms which can absorb radiation and reduces the analytical signal. Analytical errors can occur when the samples and standards exhibit different degrees of ionization. The simplest way to control ionization interferences is to add a large excess of an easily ionizable element such as potassium to both the samples and the standards. The electrons provided by the ionized potassium combine with the ions of the element being determined and increase the number of atoms which can absorb radiation.

Another potential interference, background absorption, occurs when high salt concentrations in the sample cause either molecular absorption or the scattering of hollow cathode radiation in the flame. Correction is usually made with an accessory which measures the molecular absorption and light scattering and then automatically corrects the analytical signal for this interference. These accessories, called background correctors, are purchased with most atomic absorption instruments used for water analysis.

SOLVENT EXTRACTION

Although the determination of many major elements in water is easily done with little sample preparation, some elements are present at concentrations below the normal range of the instrument. In these cases, various techniques are used to increase the concentration of the elements being determined in the sample. Occasionally the sample may be simply evaporated to a smaller volume before analysis, but this procedure is time-consuming if a large number of samples must be processed and may bring on problems of matrix or background interferences.

A frequently used method of sample concentration is solvent extraction. This involves mixing an aqueous sample with an organic chemical capable of chemically combining with the element being determined. The complex formed is then extracted into a small volume of an organic solvent and the element is thus removed from a large volume of original sample into a small volume of organic solvent, effectively concentrating the element before determination.

Solvent extraction procedures have several benefits for atomic absorption. Not only is the element being determined concentrated, but if the correct organic solvent is chosen, the inherent analytical sensitivity may be increased several-fold merely by the presence of the samples in the organic solvent instead of in water. For matrices such as seawater that would create interference problems if aspirated directly, solvent extraction also provides a means of removing the element from a complex solution to a simpler one.

Generally, solvent extraction procedures are designed to extract several metals simultaneously so they may all be determined in the same organic solvent solution. The most commonly used chelating agent is ammonium pyrrolidinedithiocarbamate, understandably called "APDC." APDC chelates most of the toxic heavy metals at the same time. Lead, cadmium, copper, zinc, iron, silver, nickel, cobalt and others may be extracted at a pH of 2 to 6. Methyl isobutyl ketone, or MIBK, is the solvent most often used in atomic absorption analysis. For metals such as aluminum or beryllium another chelating agent, 8-hydroxyquinoline, may be used. Occasionally mixtures of several chelating agents are used for maximum efficiency in extraction.

GRAPHITE FURNACE ATOMIZATION

Although the flame is the most widely used and most convenient means of transforming the elements in solution to the free atomic state, it has several drawbacks. Even modern burner systems make inefficient use of the sample. Only the smallest droplets formed in the burner chamber actually get to the flame. The remainder are wasted and run down the drain. This, of course, reduces the sensitivity from what it might otherwise be. In addition, the flame gases move the sample atoms through the light path in a very small fraction of a second, giving them only a minimal opportunity to absorb light from the source lamp and contribute to the analytical signal.

In order to remedy these faults and utilize more efficient sampling, flameless atomization systems have been devised. The most popular of these, the Graphite Furnace, has sensitivities and detection limits many times lower than the flame. The system is built around a small electrically heated graphite tube situated in the light path in place of the burner system. A small volume of sample, generally 5 to 100 μl, is pipetted into the Graphite Furnace. The furnace power supply is then programmed to pass increasing amounts of current through the graphite tube to successfully dry the sample, char off any organic material and finally atomize the sample into the light path. During atomization, temperatures as high as 2800°C are attained within the Graphite Furnace. During drying and charring stages of analysis, water vapor and vaporized organic material are swept out of the tube by a high flow of an inert gas such as nitrogen or argon. During atomization, the gas flows are automatically reduced to allow the sample atoms to stay in the light path for several seconds and thus give a larger absorption signal. The graphite tube is water-cooled, allowing it to return from its maximum temperature to room temperature in less than 30 seconds. A typical analysis takes about 90 seconds, including furnace cooldown time.

The use of all the sample and the long residence time in the light path greatly enhances the sensitivity of the Graphite Furnace over that of the flame. Typically, detection limits are improved by a factor of 100-fold or more by using the furnace as an atomization system. Table I shows comparative detection limits for some elements for the flame and for the Graphite Furnace. We see that for some of the most important toxic elements such as arsenic, cadmium, lead and selenium, the Graphite Furnace offers much improved detection limits. Many water laboratories are switching to the furnace for their low-level analyses to avoid time-consuming solvent extraction procedures.

A great deal of attention has been given to the Graphite Furnace lately for the determination of such elements as arsenic and selenium. These elements have poor detection limits by conventional flame atomic absorption techniques, and existing colorimetric techniques are difficult and time consuming. Recently, the U.S. Environmental Protection Agency developed a method utilizing the Graphite Furnace for the determination of selenium in waters, sediments and sludges (Martin *et al.,* 1975). The procedure involves a preliminary nitric acid-hydrogen peroxide digestion to break down any volatile organic selenium compounds. Since selenium is known to be quite volatile, it is stabilized by the addition of a nickel salt to the sample before injection into the Graphite Furnace. The nickel-selenium compound formed allows charring the sample at high temperatures in the furnace before atomization. Detection limits of 0.0002 to 0.002 mg/l are obtained, dependent upon the sample type and procedure used.

Table I. Selected Detection Limits,[a] (mg/l)

Element	Flame Atomization	Graphite Furnace[b]
Aluminum	0.02	0.00005
Arsenic[c]	0.2	0.00015
Cadmium	0.001	0.000003
Calcium	0.0005	0.00005
Chromium	0.003	0.00001
Copper	0.002	0.00002
Iron	0.005	0.00002
Lead	0.01	0.00005
Magnesium	0.0001	0.000004
Manganese	0.002	0.00001
Mercury[d]	0.25	0.0003
Nickel	0.005	0.0002
Potassium	0.002	0.00002
Selenium[c]	0.2	0.0005
Silicon	0.02	0.0002
Sodium	0.0002	0.00005
Zinc	0.001	0.000001

[a]Obtained with Perkin-Elmer Model 603.
[b]Obtained using 100-μl sample volume.
[c]Detection limit 0.00015 mg/l using hydride generation.
[d]Detection limit 0.00005 mg/l using flameless mercury technique.

The absorption signals generated by the Graphite Furnace are peak-shaped and very rapid. A typical peak signal takes only one or two seconds. Signals this rapid may be displayed in several ways. A recorder is most generally used, but modern instruments have the capability to display either the peak height or peak area directly on the spectrophotometer digital display or accessory printer. With the newer instruments calibration with the Graphite Furnace is as easy as with the flame, and many flameless analyses are becoming quite routine. Determinations using this flameless technique, while still not quite as straightforward or convenient as the flame, have now become commonplace in the water laboratory, because of their extreme sensitivity. Their routine use has been made even easier by the recent availability of an automated sampling system for the Graphite Furnace.

HYDRIDE GENERATION SYSTEM

For the determination of very low levels of arsenic, selenium, bismuth and antimony, an accessory sampling device—the hydride generation system—can be utilized. These elements from volatile compounds are called covalent hydrides under suitable chemical conditions. The sample to be analyzed is treated with sodium borohydride to form the volatile hydrides. These gases are then trapped and collected in an expandible balloon reservoir until the chemical reaction has stopped. The gases are then released and passed into an argon-hydrogen flame where the gaseous compounds are broken down and the atomic absorption of the chosen element is measured and the resultant signal is displayed on a recorder or directly on the spectrophotometer digits.

This technique has resulted in arsenic and selenium detection limits of less than 0.0002 mg/l when sample volumes of 20 ml are used. Since all the arsenic or selenium from a large volume of sample is introduced into the flame at once, the sensitivity is greatly enhanced over that of the routine air-acetylene flame.

Hydride generation is one of the commonly accepted detection methods for arsenic and selenium in water samples. When compared with the Graphite Furnace methods for these elements, hydride generation has advantages and disadvantages. Its biggest plus is cost, being in the $500 range for the accessory system compared with over $4000 for the Graphite Furnace. Disadvantages include the fact that it is applicable to only several of the elements required for water quality analysis, whereas the Graphite Furnace is useful for at least 40 different elements. Both techniques demand more sample manipulation than the flame and are known to be susceptible to interferences from certain sample types. Similar solution detection limits are obtained for arsenic and selenium with either method, although the hydride generation procedure calls for 20 ml of solution while the Graphite Furnace requires only 100 μl to obtain the published detection limits. If costs were not a factor, the Graphite Furnace would probably be the method of choice for most laboratories because of its greater flexibility.

FLAMELESS MERCURY SYSTEM

During the past several years, concern over the possibility of widespread mercury pollution in the environment has generated an intense interest in the measurement of mercury at very low levels. For pollution monitoring, the need exists for routine mercury determinations down to part per billion concentrations. With conventional flame sampling methods, the detection limit for mercury is about 0.2 mg/l. Fortunately, mercury is unique in that elemental mercury has appreciable vapor pressure at room temperature. Therefore, it is possible to determine mercury without the use of the conventional burner system required for atomic absorption measurements. Several highly sensitive flameless systems for the determination of mercury have been recently developed. With these systems, the mercury in the sample is reduced chemically to its elemental state. The sample is then swept, as a vapor, through an absorption cell placed in the sample beam of an atomic absorption instrument. The cell replaces the flame normally present.

For the determination of mercury in natural water, the analytical procedure used with the flameless mercury system is as follows: the sample is placed into the reaction vessel and the volume made up to approximately 100 ml with deionized water. A few drops of potassium permanganate are added to oxidize the sample. Residual organics are destroyed by adding 4-5 ml of concentrated sulfuric and nitric acids. The excess permanganate is removed by adding 4-5 ml of hydroxylamine hydrochloride. Finally, stannous chloride is added to reduce the mercury to its elemental state. The reaction vessel is then immediately connected to the mercury analysis system. After connecting the reaction vessel, a small air pump is switched on. The pump blows air into the sample solution through a fritted glass bubbler, which produces an even distribution. The air, carrying the mercury vapor, passes through a drying agent to remove water vapor, into a 15-cm absorption cell, and then in a closed cycle back into the air pump. The cell, made of plastic with detachable windows, is mounted on a bracket which fits into the neck of the burner mixing chamber.

With the flameless mercury system, an absorption equilibrium is established in about 30 seconds, after which the absorbance is measured. By preparing additional samples while the first sample is connected to the system, one operator can analyze about 30 samples per hour. When the determination is complete the operator shifts a clamp which permits the gas stream to pass through a scrubber. The activated charcoal in the scrubber quickly absorbs the remaining mercury in the system, thus ensuring operator safety.

The flameless mercury system provides accurate measurements even at concentrations much lower than 0.001 mg/l. For the analysis of solid materials, the sample must first be dissolved with oxidizing acids. Procedures have been developed for a wide variety of samples, including biologicals such as blood, urine, tissue, brines and waters,and also rock and soil samples. An additional feature of the flameless mercury system is that it can be removed easily and quickly. If the operator wishes to determine other elements by conventional flame sampling, the flameless mercury system can be removed and the burner installed in 2-3 minutes.

CONCLUSION

Atomic absorption appears to be one of the most versatile and valuable tools available to the water chemist. Not only does it allow him to do the major elements such as calcium and magnesium much more easily and rapidly than present colorimetric or titrimetric methods, but it allows the determination of tricky toxic elements at low concentrations. Whether it is cadmium at 0.001 mg/l with the Graphite Furnace, mercury at 0.0001 mg/l with the flameless mercury system or zinc at 0.01 mg/l with the flame, it is atomic absorption that makes the determination possible, This technique is no longer just a reference method for the regulatory agencies, but a routine tool for the small water lab, be it in a water treatment plant, municipal sanitation district or in industry. Perhaps this brief discussion of atomic absorption as it applies to water analysis will aid in the quest for better analytical capabilities and ultimately better water quality for all of us.

REFERENCES

Brown, E., M. W. Skougstad and M. J. Fishman. *Methods for Collection and Analysis of Water Samples and Gases,* Book 5, Chapter A1 (Washington, D.C.: U.S. Government Printing Office, 1970).

Martin, T. D., J. F. Kopp and R. D. Ediger. *Atomic Absorption Newsletter* 14:109 (1975).

U.S. Environmental Protection Agency. *Methods for Chemical Analysis of Waters and Wastes* (Washington, D.C.: Office of Technology Transfer, 1974).

11

TREATMENT OF WATER AND WASTEWATER FOR REMOVAL OF HEAVY METALS

Roy F. Weston, P. E., Chairman,
and Robert A. Morrell, P. E.

WESTON Environmental Consultants-Designers
West Chester, Pennsylvania

INTRODUCTION

Pollution of our waters by heavy metals has been receiving increased attention because of the toxicity to individual living organisms, most importantly to human beings. For many years, water treatment plants have been concerned with the levels of manganese and iron in water supplies because of the taste and color they impart to the water. More recently, health effects of heavy metals, rather than these aesthetic effects, are becoming a topic of concern. Table I summarizes the U. S. Public Health Service Drinking Water Standards for heavy metals. Most of these heavy metals have been found in levels close to or exceeding these standards in distribution systems throughout the country (Culp and Culp, 1974).[1]

In 1968, the American Water Works Association adopted water quality goals that in some instances were more stringent than the USPHS Drinking Water Standards. These goals were intended to be more exacting than these Standards with respect to aesthetic qualities. AWWA water quality goals for heavy metals are also summarized in Table I.

The importance that has been placed on the potential adverse impact of heavy metals on the environment is illustrated by the low permissible ambient concentrations in natural waters and the low effluent standards promulgated by regulatory agencies. Typical ambient- and effluent-limiting concentrations are shown in Table II.

Table I. Water Quality Goals and Drinking Water Standards for Heavy Metals

Heavy Metal	USPH Drinking Water Standards (1962) Mandatory Requirements mg/l	USPH Drinking Water Standards (1962) Recommended Requirements mg/l	AWWA Water Quality Goals (1968) mg/l
Aluminum (Al)	-	-	0.05
Arsenic (As)	0.05	0.01	-
Barium (Ba)	1.0	-	-
Cadmium (Cd)	0.01	-	-
Chromium (hexavalent)	0.05	-	-
Chromium (trivalent)	-	-	-
Copper (Cu)	-	1.0	0.2
Iron (Fe)	-	0.3	0.05
Lead (Pb)	0.05	-	-
Manganese (Mn)	-	0.05	0.01
Mercury (Hg)	-	-	-
Nickel (Ni)	-	-	-
Selenium (Se)	0.01	-	-
Silver (Ag)	0.05	-	-
Zinc (Zn)	-	5.0	1.0

In spite of the critical nature of heavy metals in drinking waters, the removal of heavy metals from drinking water supplies has been limited in practice (at least consciously) to removal of iron and manganese for aesthetic reasons. Other heavy metal removal technology has thus been applied almost exclusively in wastewater treatment. Many industries practice heavy metal removal as pretreatment prior to discharge to biological treatment systems, to avoid upsets due to the toxicity of heavy metals to bacteria.

This chapter focuses primarily on review of heavy metal removal technology, particularly as it applies to removing *low* concentrations of heavy metals. A number of industries with *high* concentrations of heavy metals in their wastewaters practice metal recovery; the methods for metal recovery are mentioned, but not emphasized in this chapter. The treatment methods covered include: (a) precipitation, (b) ion exchange, (c) adsorption, and (d) oxidation/reduction. Since solids removal is a significant factor in effective heavy metals removal, settling and filtration are also covered.

Table II. Heavy Metal Water Quality Standards[a]

Heavy Metal	Ambient Water Quality Standard (mg/l)	Effluent Standard (mg/l)
Aluminum (Al)	–	–
Arsenic (Ar)	1.0	0.25
Barium (Ba)	5.0	2.0
Cadmium (Cd)	0.05	0.15
Chromium (Cr^{+6})	0.05	0.30
Chromium (Cr^{+3})	1.0	1.0
Copper (Cu)	0.02	1.0
Iron (Fe)	1.0	2.0
Lead (Pb)	0.10	0.10
Manganese (Mn)	1.0	1.0
Mercury (Hg)	0.0005	0.0005
Nickel (Ni)	1.0	1.0
Selenium (Se)	1.0	1.0
Silver (Ag)	0.0005	0.10
Zinc (Zn)	1.0	1.0

[a]Current Standards of the State of Illinois. Changes recommended by an Effluent Standards Advisory Group include: lower chromium (Cr^{+6}) from 0.30 to 0.10; lower copper (Cu) from 1.0 to 0.50; raise lead (Pb) from 0.10 to 0.20; raise mercury (Hg) from 0.0005 to 0.003; and keep selenium (Se) as an effluent standard rather than an ambient water quality standard.

PRECIPITATION

Simple precipitation is the oldest and most widely used method for removal of heavy metals. It is also a very effective and well-proven method, and will probably continue to be the most popular for removing heavy metals, even to very low concentrations. When designed and operated properly, precipitation methods for removing heavy metals are very effective.

Most metal hydroxides are relatively insoluble in water. Their precipitation is governed by the relative concentrations of the precipitation chemical and of the metal ion in solution, and by the pH. An excess of precipitation chemical beyond the amount needed to meet the stoichiometric relationship is required. This excess can best be determined from practical experience. In most cases, the metal concentration in solution in the effluent is a function of the final chemical equilibrium treatment condition, and is independent of the initial metal concentration. Generally, as the pH increases, the solubility of the metal hydroxide decreases

(Figure 1). While heavy metal precipitation generally depends on this metal hydroxide solubility, other precipitates (*e.g.*, metal oxides and sulfides) are also important in some cases.

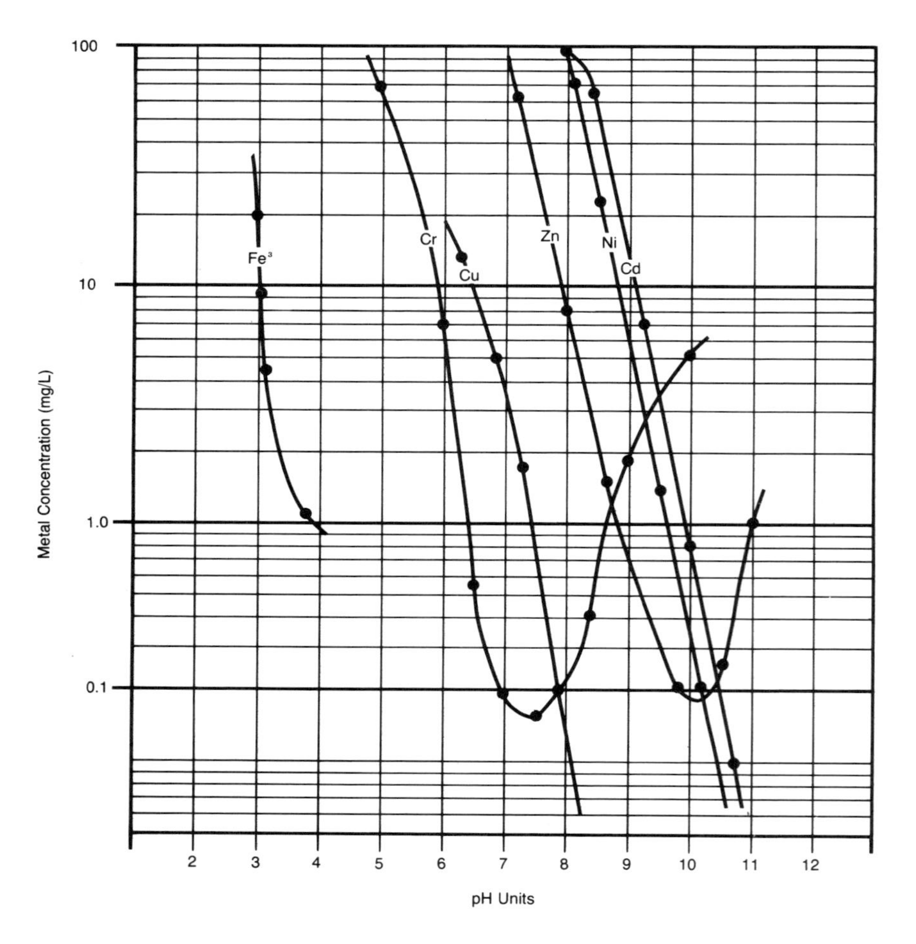

Figure 1. Precipitation as hydroxide salts.

Since many of the heavy metals form insoluble hydroxides or oxides at a pH of 11, lime treatment is effective in the precipitation of these metals. To remove the precipitates, lime treatment must be followed by settling, and to achieve low concentrations of heavy metals in the effluent, the settling must often be preceded by coagulation and followed by filtration. Table III summarizes the extent of heavy metals treatment

Table III. Heavy Metal Removal by Precipitation

	Lime Treatment		Other Precipitation Methods	
Metal	pH	Effluent Concentration (mg/l)	Concentration (mg/l)	Description
Arsenic (As)	11.0	0.03	0.05[a]	Sulfide @ pH 6.7
			0.05[a]	Ferric chloride coagulation
Barium (Ba)	11.5[a]	0.94[a]	0.03[a]	Ferric sulfate with lime @ pH ≈ 10.
	11.0[b]	1.3[b]	0.27[a]	Ferric sulfate @ pH 6
Cadmium (Cd)	10.0[a]	0.1[a]	0.04[a]	$Fe(OH)_2$ & lime @ pH 10
	11.5[a]	0.014[a]	0.05[a]	$Fe(OH)_2$
Chromium (Cr) (hexavalent)		Cannot be precipitated		
Chromium (Cr) (trivalent)	9.5[b]	0.4[b]		
	8.5-9.5[a]	0.06[a]		
Copper (Cu)	9-10.3[a]	0.5[a]		
Iron (Fe)	10.8[b]	0.1[b]	0.5[a]	Oxidation of ferrous to ferric results in precipitation of hydroxide at pH 7
Lead (Pb)	11.5[a]	0.02[a]	0.03	Ferric sulfate and lime @ pH 10
			0.03	Ferric sulfate @ pH 6.0
Manganese (Mn)	10.8[b]	<0.1	0.05	Oxidation of manganous results in precipitation of hydroxide at pH 7
Mercury (Hg)	Not applicable		0.1-0.3	Sulfide after oxidation to mercuric ion
Nickel (Ni)	11.5[a]	0.1-0.2[a]	0.35[a]	Ferrous sulfate and lime @ pH 10
Selenium (Se)	Not applicable		0.5[a] (little supporting data)	Sulfide at pH 6.5 (little supporting data)
Silver (Ag)	11.0	0.4[a]	1.4[a]	Chloride
			0.1[a]	Lime and ozone @ pH 10
			?	Sulfide
Zinc (Zn)	?	1.0[a]	very low[a]	Sulfide @ pH 2
	9.5	0.3[b]		

[a]From Patterson (1975).
[b]From Culp and Culp (1974).

achievable by lime treatment, as well as other precipitation methods. A review of this table makes it clear why lime treatment is so important for heavy metals removal. It is a very effective technology.

Many water treatment plants practice heavy metal removal incidentally, in the course of lime-soda water softening. Even without the lime-soda process, water treatment plants practice some degree of heavy metal removal merely through solids removal (*i.e.*, coagulation, settling and filtration), because many heavy metals are relatively insoluble even at neutral pH.

The effectiveness of heavy metal precipitation can be greatly reduced by interferences and/or a number of complexing agents. Although various organic materials can form complexes with metals, the most common heavy metal complexing agent is cyanide, which complexes with many metals, thus increasing the overall solubility of the metals. Removal of cyanide is usually required for effective precipitation of heavy metals. The most common method for such removal is oxidation by chlorination to carbon dioxide and nitrogen.

Precipitation processes are not the complete answer to all heavy metal removal problems, but are often at least part of the answer. Precipitation is a common process in water treatment plants, which for years have been involved with unit processes such as lime softening, flocculation, sedimentation, aeration and filtration. Thus, many water treatment plants have been achieving some heavy metals removal, and with some process modification, greater removals may be achieved with existing equipment.

SOLIDS REMOVAL

Effective solids removal is extremely important in heavy metals removal by precipitation. In removing low concentrations of heavy metals, solids removal, rather than solubility, often becomes the limiting factor. Metal hydroxides, although insoluble, tend to form bulky but light flocculent particles that often are difficult to remove by clarification unless coagulant aids are used.

Filtration following clarification is usually required to achieve very low concentrations of heavy metals. Good examples of the importance of filtration are shown in Table IV, which indicates that filtration makes as much as one order of magnitude difference in the concentration of heavy metal achieved after precipitation and clarification.

Another important role of solids removal is as a pretreatment operation in heavy metal removal by processes such as ion exchange, reverse osmosis, activated carbon adsorption and electrodialysis. Accumulation of solids in reverse osmosis and electrodialysis membranes or in ion exchange and carbon media can adversely affect the operation of these processes.

Table IV. Attainable Concentration of Various Metals

Metal	Precipitation and Clarification (mg/l)	With Filtration (mg/l)	Reference
Lead	0.2	0.019	Patterson, 1975
	0.25	0.029	Patterson, 1975
	0.25	0.03	Patterson, 1975
Chromium (Trivalent)	2.7	0.63	Culp and Culp, 1974
Copper	0.79	0.32	Culp and Culp, 1974
Selenium	0.0103	0.00932	Culp and Culp, 1974
Zinc	0.97	0.23	Culp and Culp, 1974

In summary, efficient solids removal is invariably required to achieve low heavy metal concentrations, and filtration is usually required.

ION EXCHANGE

The ion exchange process has been used by many industries for water treatment when an extremely high purity of water is required. Ion exchange, however, is capable of removing only ionic species from water; suspended materials (solids) are not removed by ion exchange and are usually detrimental to the process, because they can foul the ion exchange beds.

Ion exchangers are simply insoluble electrolyte materials that exchange ions with a solution. There are two main classes: cation exchangers and anion exchangers. A cation exchanger removes only positively charged ions from solution, while an anion exchanger removes only negatively charged ions from solution. Ion exchange processes are very effective in removing heavy metals to very low concentrations, but they are relatively nonselective and also remove other ions of like charge. Thus, if the removal of heavy metals in the presence of high concentrations of other dissolved inorganics (*e.g.*, Na^+, Ca^{++}) is desired, ion exchange will not selectively remove the heavy metals, and the resin will be spent rapidly.

Ion exchange resins are usually regenerated, and the spent regenerant is, in effect, a more concentrated wastewater stream contaminated with the same ions (heavy metals) that were removed from the more dilute, treated stream. This wastewater must be disposed of, and the disposal often requires treatment for removal of heavy metals. In effect, ion exchange will remove low concentrations of heavy metals very effectively and concentrate them in a stream of less volume, which can be treated by precipitation or other recovery processes. Ion exchange, therefore, is most applicable as a scavenging or polishing treatment unit.

The utilization of ion exchange in water treatment has been mainly in connection with water softening, and a strong acid cation exchanger (sulfonated copolymer resin of styrene and divinylbenzene) is most often used. The ion exchange reactions can be represented as follows:

$$2RSO_3\ Na + Ca^{++}\ (\text{or } Mg^{++}) = (RSO_3)_2\ Ca\ (\text{or } Mg) + 2Na^{+}$$

A 10% (or stronger) solution of NaCl is normally used to regenerate the resin.

Most heavy metal cations will exchange with the strong acid cation resins used in water softening. As mentioned previously, selectivity for removal of specific ions is not good, but there are conditions where certain ions are more readily exchanged. Concentration and valence are two factors that influence cation exchange. Ions present in high concentrations will exchange more than those in low concentrations. At equal concentration the removal of divalent cations (*e.g.*, Mn^{++}, Cu^{++}, Pb^{++}) will be greater than that of monovalent cations (*e.g.*, Na^{+}, Li^{+}, $NH_4{}^{+}$). A consequence of this valency influence is the increase of sodium concentration in the treated water.

ADSORPTION

The use of adsorption for removal of heavy metals has been reported for arsenic, cadmium, chromium, copper, mercury and nickel. The mechanism for metals removal with adsorptive materials such as activated carbon is not fully understood, but the following can be postulated with reasonable confidence:

1. Heavy metals are known to form soluble complexes with organic compounds. In the presence of such organics, heavy metals removal may be achieved by complexing followed by adsorption of the organics.
2. Heavy metals can form hydroxide complexes which, in effect, can act like polymers. It is possible that these hydroxide complexes can form hydrated molecules large enough for adsorption to be effective.

Generally speaking, adsorption processes are not as applicable as precipitation or ion exchange; however, their use should be considered in special applications, particularly those in which precipitation and ion exchange are ineffective.

OXIDATION/REDUCTION

Oxidation/reduction processes play an important role in heavy metals removal, particularly by precipitation. For example, to achieve effective precipitation of iron at a near neutral pH, ferrous iron must be oxidized to ferric, which occurs very readily at neutral pH in the presence of oxygen. Aeration is usually sufficient to accomplish this oxidation.

Manganese is soluble in water in the forms of manganous and permanganate ions. The permanganate ion is a strong oxidant and is reduced under normal circumstances to insoluble manganese dioxide. The manganous ion, however, must be oxidized to the insoluble manganic ion. Unlike iron, the manganous ion is not oxidized readily by means of aeration at neutral pH, and requires either aeration at a high pH ($\approx$ 10) or chemical treatment. Chemical treatment involves the use of a strong oxidant such as chlorine, ozone, hypochlorite, chlorine dioxide, manganese dioxide or potassium permanganate.

To precipitate mercury as mercuric sulfide, mercurous and organic mercury compounds must be oxidized to mercuric ion. The reduction of mercury ions to free elemental (insoluble) mercury has also been proposed as a method of mercury removal by precipitation.

As indicated in Table III, trivalent chromium can be precipitated as a hydroxide by means of lime treatment but hexavalent chromium cannot. Reduction of hexavalent chromium from a valence state of +6 to +3, and subsequent precipitation of the trivalent chromium ion, is the most common method of hexavalent chromium removal. The most common reduction process is an acid reduction in which the pH is lowered with sulfuric acid to a pH of 3 or below, and the hexavalent chromium is converted to trivalent chromium with a chemical reducing agent such as sulfur. Other reducing agents include sodium bisulfite, sodium metabisulfite, sodium hydrosulfite and ferrous sulfate.

A very common complication to heavy metal precipitation is the presence of cyanide, a toxic contaminant in its own right, which is often found in wastewater streams with heavy metal contamination. The cyanide form complexes with heavy metals, thus increasing the solubility of the metals and decreasing the effectiveness of precipitation. Cyanide is an organic structure which can be destroyed by oxidation to carbon dioxide and nitrogen, and the most common oxidant used for its destruction is chlorine. Complete oxidation of cyanide is usually a two-step procedure requiring close control of pH. The first step is oxidation of the cyanide to cyanate at pH 10 or higher. The second step is oxidation of cyanate to CO_2 and nitrogen by addition of excess chlorine at a pH of 8-8.5. Cyanate can also be oxidized to CO_2 and ammonia by acid hydrolysis at pH 2-3, usually by the addition of sulfuric acid.

MISCELLANEOUS PROCESSES

A number of other treatment processes are applicable for removal of heavy metals, particularly in specialized applications. Reverse osmosis, electrodialysis and evaporation processes have been used to achieve concentrations of heavy metals for recovery purposes. Additional treatment is usually required in conjunction with these processes, however, and costs are usually quite high. Other processes that have been considered for removing heavy metals from water include solvent extraction and freezing. Generally speaking, these processes warrant consideration only where recovery of a valuable metal is practical.

SLUDGE DISPOSAL

The end result of most heavy metal removal processes is a sludge, which ultimately must be disposed of. Typically, heavy metal sludges are landfilled. An important consideration in the disposal of heavy metal sludges is that many of the solids can go back into solution when the pH decreases. The reversibility of precipitation is such that rainwater, with its relatively low pH, can cause the heavy metal solids to go back into solution. Treatment of landfill leachates has been proposed, but this merely results in more heavy metal sludge and, thus, a cyclic operation.

Landfilling of heavy metal sludges is indeed a feasible method of disposal, if proper care is taken to segregate it from other sludges and prevent contact with surface or ground water. Nevertheless, the trend towards more frequent occurrence of leachate problems points to the need and possible future trend towards heavy metal source control and metal recovery.

INTEGRATED APPROACH–CASE STUDY (PATTERSON, 1976)

To illustrate some of the principles touched on in this chapter, a case study of industry discharging metal wastes is presented. This case study is based on a Weston industrial client, and was previously presented at AIChE's 82nd National Meeting (Patterson, 1976). Although this particular case study involves the removal of relatively high metal concentrations from water, the principles involved are nevertheless applicable to removal of low concentrations.

The treatment facilities for this industry were designed on the basis of the influent and effluent waste characteristics shown in Table V; the effluent quality predicted in Table V was based on wastewater treatability studies. A flow diagram of the treatment process involved is shown in Figure 2.

Table V. Design Influent and Effluent Characteristics[a]

	Influent	Effluent
Total flow (gpd)	274,000	274,000
Cyanide flow (gpd)	12,500	12,500
Chromium flow (gpd)	11,500	11,500
COD (mg/l)	300	160 (100)[b]
Suspended solids (mg/l)	195	10
pH	10.5	8.5
Total dissolved solids (mg/l)	740	1,025
Copper (mg/l)	29	2.65
Zinc (mg/l)	5.0	0.50
Cadmium (mg/l)	1.8	0.07
Nickel (mg/l)	1.5	0.01
Total chromium (mg/l)[c]	0.16	0.05
Total heavy metals (mg/l)	37.5	3.28
Cyanide (mg/l)	200	0.05

[a]From Patterson (1976).
[b]Permit application stated 100 mg/l average, 150 mg/l maximum. (Actual permit did not include any limitations.)
[c]Based on data from existing plant.

The heart of the process is single-stage lime treatment, a very traditional treatment process. Because of the cyanide and hexavalent chromium present in the wastewaters, the more concentrated streams are isolated and pretreated (through cyanide oxidation and chromium reduction) prior to the lime treatment.

The importance of good solids removal is well-reflected in the design of this treatment system, which includes a flocculator-clarifier followed by polishing filters and effluent polishing lagoons. Since a number of heavy metals required precipitation in a single stage, design of the process required knowledge of the effect of pH on solubility. Investigation disclosed that this relationship varied from time to time, and Figure 3 illustrates the typical soluble metal concentration vs pH data for this wastewater at a given time. The minimum concentration for each metal varies from day to day, depending on many production variables and probably also on the presence of complexing agents. Figure 3 indicates that the optimum pH for Zn removal is 8-9, while for copper it is nearly 11. To achieve maximum overall removal of metals, the system was designed to be maintained at a pH of 9-10. Should lower concentrations of metals in the effluent be required, a two- or three-stage precipitation process would be required, because the minimum solubility of each of the metals involved occurs at a different pH.

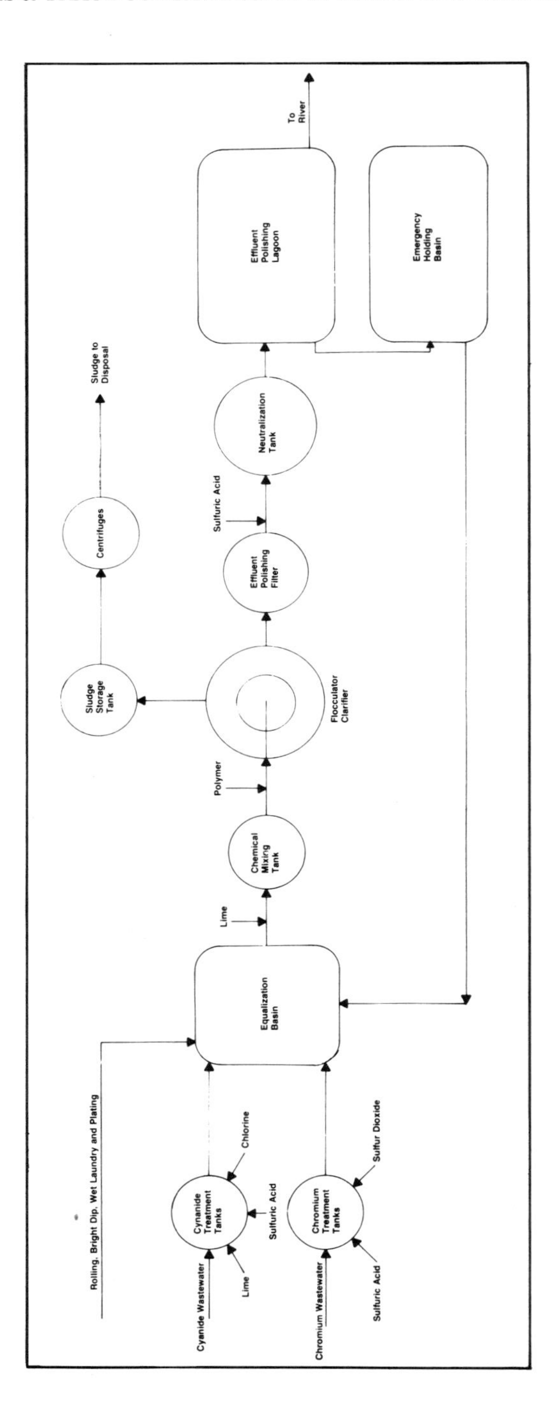

Figure 2. Simplified process flow diagram.

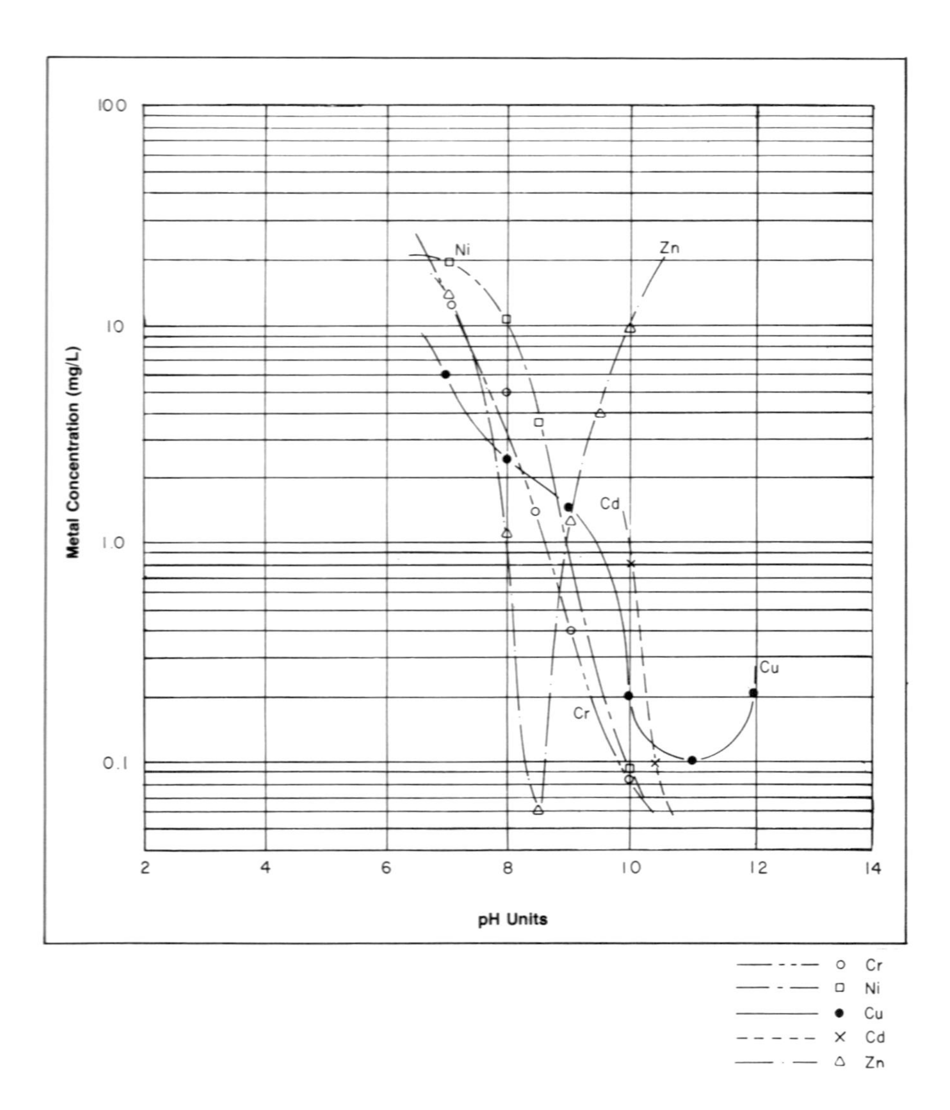

Figure 3. Soluble metal concentration vs pH.

Table VI is a summary of the operating records from the plant during the twelve months following start-up. The plant is operated to minimize total metals in the effluent, and an onsite atomic absorption unit provides current operational data needed for pH adjustment and optimal plant control.

Table VI. Average Monthly Effluent Quality[a]

Month	Flow (mgd)	COD (mg/l)	SS (mg/l)	Cyanide (mg/l)	Cadmium (mg/l)	Total Chromium (mg/l)	Copper (mg/l)	Nickel (mg/l)	Zinc (mg/l)	Total Metals (mg/l)	Remarks
December 1973	0.235	62	8	0.33	0.05	0.03	2.33	0.44	0.78	3.61	
January 1974	0.218	72	5	0.12	0.04	0.03	2.41	0.56	0.42	3.46	
February 1974	0.210	66	7	0.04	0.02	0.02	0.69	0.50	0.18	1.40	Isolated stripper problem
March 1974	0.271	61	5	0.03	0.01	0.02	0.45	0.22	0.24	0.94	Represents optimum single-stage plant operation
April 1974	0.255	62	14	0.03	0.0	0.02	0.46	0.17	0.16	0.80	
May 1974	0.219	58	11	0.02	0.01	0.02	0.59	0.17	0.16	0.95	
June 1974	0.251	79	24	0.03	0.01	0.02	0.41	0.21	0.01	0.75	
July 1974	0.190	81	10	0.02	0.01	0.02	0.39	0.23	0.11	0.76	
August 1974	0.253	78	22	0.02	0.01	0.02	0.51	0.26	0.26	1.06	Reduced operating pH
September 1974	0.266	71	19	0.04	0.01	0.02	0.59	0.30	0.25	1.17	Conserve chemical cost
October 1974	0.283	65	17	0.03	0.01	0.02	0.58	0.45	0.38	1.38	
November 1974	0.273	45	2	0.04	0.01	0.02	0.77	0.43	0.44	1.67	

[a]From Patterson (1976).

The copper problem encountered in this project is of particular interest. Unusually high copper concentrations after lime treatment were often observed, both in the treatability testing program (see Table V, which predicts an effluent copper concentration of 2.65 mg/l) and in the first two months of operation of the treatment plant (2.33 mg/l in December 1973 and 2.41 mg/l in January 1974). Investigation of the problem disclosed that the cause was intermittent discharges of a small quantity of organic material into the process sewer. When this discharge was stopped, the concentration of copper in the effluent decreased substantially (0.39 to 0.77 mg/l, averaging 0.54 mg/l in the next 10 months). Apparently, the organic material had been complexing with the copper, thereby increasing its solubility and inhibiting its precipitation.

SUMMARY AND CONCLUSIONS

Heavy metals removal has historically relied on precipitation and good solids removal by sedimentation and filtration. However, a number of other treatment processes, most notably ion exchange, can also accomplish heavy metals removal. Other processes, such as adsorption, freezing, reverse osmosis, electrodialysis, evaporation and solvent extraction have limited practical applicability. Problems associated with the ultimate disposal of heavy metal wastes indicate a probable future trend toward heavy metal source control and recovery.

It is important to note that both the precipitation and the ion exchange processes, particularly precipitation, are traditional water treatment processes. Lime treatment is widely practiced for removal of calcium and magnesium ions in water softening. Ion exchange has also been used for the same purposes; however, because of higher cost, ion exchange normally is used for water softening only in specialized applications where the quality of water must be very high. It is thus likely that many water treatment plants are already achieving some degree of heavy metals removal, and that with minor process modifications high degrees of removal could be achieved.

Carbon adsorption may be important where the presence of organic compounds results in metal complexes that interfere with precipitation. Carbon adsorption also is a fairly common water treatment process.

The use of other heavy metal removal processes should not be ignored, but their applicability is highly specialized, and their use is justified only after full consideration of the more traditional heavy metal removal technology.

REFERENCES

American Water Works Association. *Water Quality and Treatment*, 3rd ed. (New York: McGraw-Hill Book Co., 1971).

Calmon, C. "Modern Ion Exchange Technology," *Ind. Waste Eng.* (April/May 1972), pp. 12-16.

Culp, G. L. and R. L. Culp. *New Concepts in Water Purification*, Environmental Engineering Series (New York: Nostrand Reinhold, 1974).

Patterson, J. W. *Wastewater Treatment Technology* (Ann Arbor, Mich.: Ann Arbor Science Publishers, Inc., 1975).

Patterson, K. E. "Traditional Vs State-of-the-Art Methods of Treating Wastewaters Containing Heavy Metals," presented at AIChE, 82nd National Meeting, Atlantic City, New Jersey (1976).

Permutit. *Proceedings of Seminar on Metal Waters Treatment Featuring the Sulfex Process* (October 1976).

12

THE ASBESTOS PROBLEM

T. J. Williams, Ph.D.
Bureau of Disease Control and Laboratory Services
Michigan Department of Public Health
Lansing, Michigan

INTRODUCTION

The subject of this chapter covers a very broad area and normally one would concentrate on a selected topic. However, because of my work with our department advisory committee and my involvement with the Attorney General's office during the litigation against Reserve Mining Company, I have had the opportunity for an excellent overview of the entire problem. I will, of course, limit myself to the asbestos problem as related to drinking water, and will attempt to cover most of what we now know about the asbestos problem, and in many cases what we do not know. My main intent is to bring an overall perspective to this problem, including where it has led us in the past and where we stand today.

We should first consider what is meant by the asbestos problem. In many respects it would seem as though this problem suddenly came to our attention just a few years ago. I am sure the general public feels this way as perhaps do some reading this book. Did we suddenly discover that significant populations were ingesting asbestos particles? No. Small amounts of asbestos have been ingested for years in areas where asbestos filters in processing and, most importantly, with the use of asbestos-cement pipe for water transmission lines. All these aspects have been previously considered, and all involved generally agreed that there was no evidence, either in the unusual occurrence of disease or from scientific experiments, that such exposure could be related to health problems. Did we suddenly find out that asbestos can be related to various diseases? No. This too has been known for many years, and formal controls on occupational exposure have long been in effect. What did happen was a decision to use possible health-related effects of asbestos in potable water supplies as a key element in the litigation to stop Reserve Mining Company's dumping of ore wastes into Lake Superior. This was done beginning June 15, 1973, with thorough publicity, which in turn generated understandable concern on the part of the many citizens involved. This action immediately and publicly raised many more questions than we had answers or, in some cases, even the means to study the proposed health hazards. Even worse, there did not appear to be adequate alternate sources of water after the announcement that a significant health hazard existed.

In many respects, this is a rather classic case of the general types of problems discussed in this book. These problems can usually be broken down into four aspects:

1. analytical difficulties;
2. definition of health hazards;
3. economical or technological feasibility of means to correct suspected problems; and
4. lawyers, politics and the news media.

As a scientist, I tend to be rather reluctant to discuss this fourth category. However, in some instances, this aspect is perhaps the biggest problem, often blocking real progress on significant problems or generating problems where none actually exist. On the other hand, these nonscientific aspects seem to provide the basis for almost all the support for environmental scientific research.

It appears that the initial announcement of asbestos occurrence in water supplies was based on X-ray diffraction data obtained at the National Water Quality Laboratory in Duluth. This method measures mass of material present versus some arbitrary standard and provides some characterization regarding mineral types (1974). This and the other methods I will discuss depend for their sensitivity on a high ration of asbestiform minerals to other suspended solids. Ironically, this finding would probably have not been possible except for the excellent quality of the water sampled with respect to total suspended solids. Initial consideration of health hazards presented a number of analytical difficulties. Most health-related studies involved specific mineral types. The general term, asbestos, seems to have a definition more closely associated with the use of the material than what it actually is. The two mineral types, chrysotile and amosite, are used for almost all asbestos products, and it appears that the general term "asbestiform" evolved from these initial considerations to mean any mineral that looks like chrysotils amosite or other fibrous amphiboles.

This, in turn, implied that analysts should be concerned with any mineral having a fiber-like structure. Trying to define a fiber in an objective manner is as difficult as trying to define asbestos. In a purely arbitrary manner, and to be on the safe side, fibers were defined as having a width-to-length ratio of 1:3 or larger, as viewed through the microscope.

Considerations related to possible health effects indicated the number of fibers present and fiber size might also be important. The fact that individual fibers could serve as the disease initiator seems to be the origin of the use of the term fibers/liter to define asbestiform content. This terminology is not very satisfying for an analytical measurement, because it has nothing to do with the amount of material present. While the mass or amount of material must remain constant, any number of factors can and apparently do change the fiber count. Initially, fibers observed were not well-defined as to size. Therefore, the early reports of up to 10 million asbestiform fibers/liter sounded impressive; however, in the conventional sense, this represented a virtually unknown amount of an uncertain substance.

It was a number of months before we began to get somewhat consistent data by a tedious, involved electron microscope technique. Roughly, this involves the following procedure.

Outline of Recommended Asbestos Analysis

1. Filter known volumes of sample.
2. Mount a known fraction of the filter (then filter removed by dissolving) or fraction of the filtered material (recovered by low-temperature ashing) on an electron microscope grid.
3. "Shadowing" of the grid is required to obtain all three particle dimensions.[a]
4. Observe particles and count those having,
   (a) fibrous appearance;
   (b) where properly oriented, characteristic electron diffraction patterns;
   (c) further particle definition, which may be obtained by particle elemental analysis using microprobe X-ray emission techniques.[a]
5. Record dimensions of all particles counted.
6. Report within each class of fibers identified,
   (a) number of fibers per liter;
   (b) dimensions (or range of dimensions) of fibers;
   (c) mass of fibers/liter (usually μg/l calculated from estimated sum of fiber volumes and given specific gravities).

---

[a]Not used routinely.

As indicated, this is an extremely expensive type of work involving on the order of $100,000 worth of equipment and operating costs of about $100 per sample.

Data, obtained by this technique and reported to the Michigan Department of Public Health (MDPH), which we consider sufficiently reliable, are shown in Table I. The results can be divided roughly into two categories by amounts reported. That is, those samples from the Silver Bay-Duluth-Superior area giving concentrations of asbestiforms in the range of 0.2-2.7 μg/l. The Silver Bay data are an excellent example of the lack of correlation between fibers/liter and microgram/liter values. This limited data also shows a curious lack of expected change in values going from plant influent to plant effluent or water in the distribution system. Values should remain constant or decrease; however, in a number of instances, values show a 2-5 fold increase in the plant effluent or distribution system as compared with plant influent. Table II shows data collected by MDPH for Michigan water supply sources. In all cases, asbestiforms were found to be less than 0.01 μg/l.

The graph in Figure 1 shows how values for asbestiform content vary over time for Duluth raw water at the Lakewood Intake. There is some variation, but with one exception the range covers less than a factor of three. Reviewing earlier arguments, I found a great deal of effort was expended attempting to link nonchrysotile amphibole content to health studies specifically involving chrysotile asbestos. This data was, therefore, more than a little surprising. It shows chrysotile to be much more predominant (again as fibers/liter) than amphibole particles.

While data such as this gives some feeling about the occurence of asbestiforms, it cannot really be applied without at least a rudimentary knowledge of related health effects. As I mentioned, there is first a question of whether various types of asbestiform minerals can be considered as having equivalent health effects. Assuming this to be true, a much more significant question is whether exposure to asbestos by ingestion only, such as with drinking water, is in any way equivalent to exposure by inhalation. The apparent relationship for the two types of exposures is indicated by the comparisons in Figure 2.

Everyone seems to have a slightly different opinion as to what all this means. The environmental studies should, of course, be pursued to find out what these relationships signify. However, the data given do not suggest a significant health risk. To the contrary, it appears to me that there is no significant health risk from drinking water containing fibers of less than 5 μ length at levels of less than 30 μg/l. Even above these levels, the effect on health remains unknown for exposure involving only ingestion.

Table I. Results of Electron Microscope Analysis (EPA Survey, 1973)[a]

Sample Location	University of California-Berkeley (mf/l)	W. C. McCrone Associates (mf/l)	(μg/l)	(fw)
Ashland, WI-LS-Plant Eff.	–	0	0	–
–Dist.	0	0	0	–
Aurora, MN-U-Plant In.	0.09	0.26	0.003	0.012
–Plant Eff.	0	–	–	–
–Dist.	0	–	–	–
Baptism R.-T	0	0	0	–
Beaver Bay-LS-Plant In.	–	0.943	1.141	1.210
–Plant Eff.	–	4.189	2.739	0.654
–Dist.	3.0	5.29	0.38	0.072
Beaver R.-T	0	0	0	–
Birch Lake Reservoir-U	0	–	–	–
Cloquet, MN-U-Plant In.	0	0.393	0.0032	0.008
–Dist.	0	–	–	–
Duluth, MN-LS-				
–Plant In.	–	2.57	0.68	0.265
–Plant Eff.	–	1.78	0.27	0.152
–Dist.	5.0	2.76	0.15	0.054
–Dist.	1.0	2.26	1.48	0.655
–Dist. Pumping Station	1.0	1.47	0.20	0.136
–Dist. Pumping Station	2.0	6.08	0.46	0.076
Eagle Harbor, MI-LS-				
–Plant In.	–	0.10	0.002	0.02
–Plant Eff.	0	–	–	–
–Dist.	–	0.105	0.0007	0.007
Eveleth, MN-U-				
–Plant In.	–	–	0.02	0.039
–Plant Eff.	–	–	0.002	0.017
–Dist.	–	–	0.004	0.014
Grand Marais, MN-LS-				
–Plant In.	–	–	0	–
–Plant Eff.	–	–	0.003	0.060
–Dist.	0	0	0.06	0.240
Hallet Well, Duluth-G	0	0	0	–
Houston, Texas- Four different system samples	0	0	0	–

Table I. Continued

Sample Location	University of California-Berkeley (mf/l)	W. C. McCrone Associates (mf/l)	W. C. McCrone Associates (μg/l)	W. C. McCrone Associates (fw)
Hoyt Lakes, MN-U-				
–Plant In.	0	–	–	–
Reserve Mining Company-T				
–Launders, west	20,000	–	–	–
–Launders, west	40,000	–	–	–
St. Louis R.-T	–	0	0	–
Stewart R.-T	–	0	0	–
Silver Bay, MN-LS				
–Plant In.	0.5	1.47	0.70	0.476
–Plant Eff.	–	10.447	0.7058	0.068
–Dist.	2.0	4.37	0.80	0.183
Superior, WI-G(?)-				
–Plant In.	0	0	0	–
–Plant Eff.	0.7	0.37	0.0132	0.036
–Dist.	0.3	1.0997	0.0661	0.060
-LS-?	0.2	–	–	–
Two Harbors, MN-LS-				
–Plant In.	–	1.10	0.24	0.218
–Plant Eff.	–	1.89	0.31	0.164
–Dist.	0.2	2.53	1.54	0.069
Virginia, MN-U-				
–Plant Eff.	–	1.1	0.04	0.036
–Dist.	–	0.94	0.03	0.032
White Pine, MI-LS-				
–Plant In.	–	0.21	0.23	1.09
–Plant Eff.	–	0.10	0.015	0.150
–Dist.	0	0.183	0.1174	0.642
Wild Rice Lake-T	–	0	0	–

[a]Source Codes: LS=Lake Superior; G=ground water; T=tributaries to Lake Superior; U=unknown.

Abbreviations: In.=influent; Eff.=effluent; Dist.=distribution system; mf/l=10^6 fibers/liter; μg/l=10^{-6} g/l; fw=average fiber mass μμg-10^{-12} g; – =samples not examined.

Table II. Results of Michigan Testing (Michigan Department of Public Health)[a]

Location	Sampling Date	Source	OM Analysis (fibers/liter)	XD Analysis (g/l)	EM Analysis
Escanaba	9/1973	Lake Michigan	ND	ND	0.31 mf/l 0.004 μg/l
Ontonagon	9/1973	Lake Superior	ND	ND	ND
Eagle Harbor	9/1973	Lake Superior	ND	ND	–
White Pine	9/1973	Lake Superior	ND	ND	–
Marquette	9/1973	Lake Superior	ND	ND	ND
Grand Rapids	11/1973	Lake Michigan	ND	ND	0.67 mf/l 0.001 μg/l
Kalamazoo	11/1973	Ground water	ND	ND	–
Van Rippers State Park	11/1973	Ground water	ND	ND	ND
Lansing	11/1973	Ground water	ND	ND	–
Pinconning	11/1973	Saginaw Bay	ND	ND	–
Detroit	11/1973	N. Detroit River	ND	ND	0.90 mf/l 0.002 μg/l
White Stone Pt.	11/1973	Lake Huron	ND	ND	ND
Monroe	11/1973	Lake Erie	ND	ND	ND
Bay City	11/1973	Saginaw Bay	ND	ND	ND
Eagle Harbor	4/1974	Lake Superior	ND	ND	ND
White Pine	4/1974	Lake Superior	ND	ND	0.19 mf/l 0.001 μg/l
Ontonagon	4/1974	Lake Superior	ND	ND	ND
Detroit SW	5/1974	S. Detroit River	ND	ND	ND
Detroit SW	5/1974	S. Detroit River (after treatment)	ND	ND	ND

[a]Abbreviations: ND=not detectable; OM=optical microscopy, not detectable, meaning asbestos-like fibers of length greater than approximately 2 μ not observed; XD=X-ray diffraction, not detectable, meaning that peaks characteristic of amphiboles or serpentine minerals not observed – estimated lower mass limit of detection 0.9 ppm; EM=electron microscopy and diffraction, not detectable, meaning that the number of fibers of length greater than approximately 0.01 μ with characteristic asbestiform properties is less than approximately 0.05 x 10^6/l, the limit of detection – means samples not examined by EM methods; mf/l=10^6 fibers/liter, μg/l=10^{-6} g/l asbestos as identified by EM techniques.

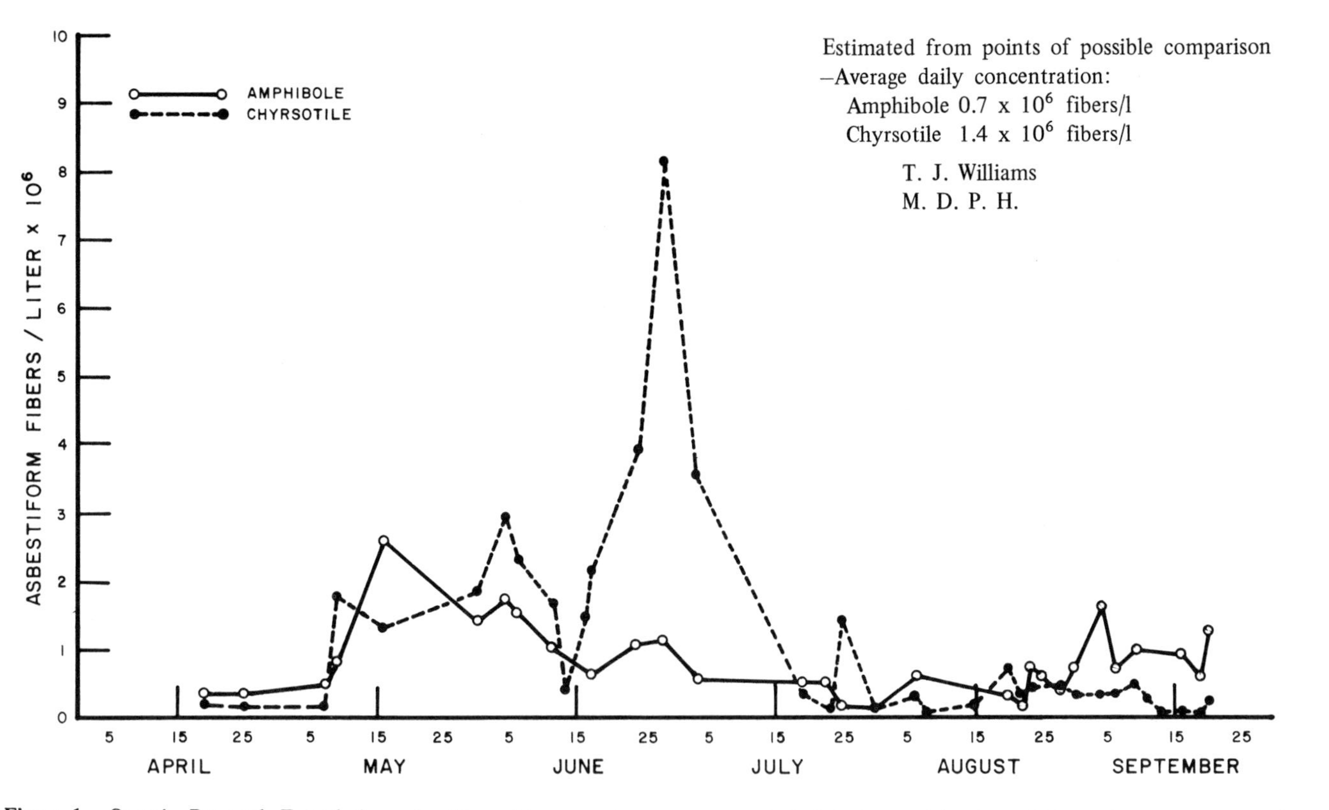

Figure 1. Ontario Research Foundation asbestiform fiber counts raw water at Duluth Lakewood intake–1974. Taken from Environmental Protection Technology Series, "Direct Filtration of Lake Superior Water for Asbestiform Fiber Removal," Appendix E, p. 8 (1975). U. S. Environmental Protection Agency Document No. EPA-670/2-75-050e.

Inhalation	Ingestion
1. Statistical studies indicate a relation between occurrence of cancer and occupational asbestos exposure. Recent studies indicate a possible relation between cancer occurrence and environmental asbestos exposure (presumably both inhalation and ingestion).	1. Statistical studies have not been done for populations exposed to asbestos by ingestion only. Subjective reports from areas such as Thetford Mines, Quebec, where asbestos occurs naturally in water supplies (up to 172 x 10^6 fibers/liter), indicate no unusual rates of disease occurrence.
2. Standards have been established for occupational asbestos air contamination.	2. Similar standards have not been set for drinking water. The approval of asbestos-cement pipe implies allowable levels of at least 0.1 μg/1 (up to 2 million fibers/liter based on data in Table I).
3. Accumulation of asbestos fibers does occur in the lungs. Penetration of lung tissue by fibers has been reported.	3. Accumulation of asbestos fibers in the intestinal tract have not been reported. Gut penetration apparently does not occur.
4. Animal experiments have shown tumors to be induced by lung exposure to asbestos.	4. Many animal feeding studies of various types of asbestos (up to 5% of diet) have shown no significant tumor formation.
5. Animal studies suggest that short asbestos fibers (less than 5 μ) are not carcinogenic.	5. The large majority of fibers reported in drinking water are less than 1 μ in length.
6. Considerations by a committee of experts selected by the AWWA Research Foundation found no evidence for increased risk of intestinal cancer at lower levels of occupational exposure studied (estimated at greater than 2.0 g total exposure over a 40-year working period).	6. A 2.0-g exposure via drinking water for an 80-year period at 2 l/day would require asbestos concentrations in drinking water of greater than 34 μg/l for the entire period.

Figure 2. Health considerations regarding asbestos exposure.

Of course, this opinion may be debated, but I have tried to base it on information now available. I am greatly concerned regarding the type of approach to these problems expressed in a communication from the Executive Office of the Council on Environmental Quality: ". . . definitive conclusions as to health risk cannot be drawn without present state of knowledge . . . [pending further study] . . . we believe action should be taken on the basis of a possible health risk." This comes close to the idea that we should take action on the basis of what we do not know, rather than what we do know. This approach perhaps has a place if the action involved has no serious consequences. However, where such action is costly or disruptive, the people involved will certainly expect objective support of this action based on agreement in the scientific-medical community. Indeed, it appears that in practice, corrective action of this nature simply cannot be accomplished without such support.

This brings us to what can be done to remove asbestiform particles from drinking water if we do decide it must be accomplished. The Duluth experience with small filters attached to sources of drinking water was not very successful. Unless routinely maintained, these devices occasionally failed putting concentrated slugs of material into water that, for the most part, is consumed as drinking water. Research done by the EPA Water Supply Research Laboratory (Figure 1) has shown good removal of amphiboles by flocculation-filtration. Results for removal of chrysotile asbestos have been erratic with much poorer overall removal. In general, it now appears that this is a problem we cannot yet solve completely with reasonably conventional water treatment techniques.

In summary, data from the Michigan Department of Public Health has been presented as completely as possible. Based on this evidence it appears that:

1. Routine monitoring of asbestos in a reliable fashion is so costly that it is beyond the means of all but the very largest of water supplies.
2. There is a need for further research regarding the estimation of risks to health from ingesting asbestos. The data presented do not, however, appear to adequately support extensive corrective action for Michigan water supplies containing asbestiforms at levels thus far encountered.
3. Proposed treatment techniques for asbestiform removal have so far been only partially effective.

It is our understanding that EPA is conducting extensive research on this subject; hopefully, we will be able to more conclusively evaluate this situation based on these additional findings as they become available.

REFERENCES

Cook, P. M., G. E. Glass and J. H. Tucker. "Asbestiform Amphibole Minerals: Detection and Measurement of High Concentration in Municipal Water Supplies," *Science* 185:853 (1974).

SECTION IV

STANDARDS

13

THE TECHNICAL AND SCIENTIFIC BASIS FOR DRINKING WATER STANDARDS

Gordon G. Robeck
Director, Water Supply Research Division
Municipal Environmental Research Laboratory
U.S. Environmental Protection Agency
Cincinnati, Ohio

For about 60 years the federal government, through the U.S. Public Health Service (USPHS), has had a responsibility to certify the safety of water supplied to and used on Interstate Carriers. The EPA now has this responsibility and is busily trying to update the present Drinking Water Standards issued in 1962. The actual process of developing criteria and setting standards traditionally has been through the combined expertise of federal staff members and nonfederal advisors. Both groups, of course, rely principally on the scientific literature, legal procedure and practical experience for selecting constituent limits. By having representation on the advisory committee from the most involved and affected public and professional groups, acceptable criteria have been developed for most constituents. Some limits, however, have a rather inadequate scientific base.

How these criteria are selected and examples of inadequate as well as adequate basis for standards are the subjects of this chapter. In 1962, for instance, the USPHS introduced standards for fluoride, nitrate and certain organics as represented by a carbon chloroform extract. Each had a much different degree of scientific and field data to support its limits, but the committee thought there was enough justification in each case to warrant inclusion. Pesticides, on the other hand, were not included for a variety of reasons.

There are many harmful substances in natural runoff as well as in man's discharges to streams, but only a few are known to occur in water consistently and at concentrations high enough to have health or aesthetic significance. Thus, a certain abundance and general widespread occurrence are necessary before a certain constituent is considered for inclusion in the Drinking Water Standards. If this policy were not followed, the analytical and enforcement chores would be unduly burdensome.

Of course, knowing the occurrence of a chemical in water implies there is an analytical method for quantifying the substance, but in reality, especially with organics, there is no quick, reasonable way to measure certain compounds within a mixture, so no sensible standard can be set for certain specific organics. Thus, analytical ability is an indirect factor in establishing enforceable standards.

Protecting human health is the primary reason for setting standards, and several considerations go into selecting health limits for pathogens or chemicals. Among these are the public health statistics related to morbidity and mortality; the population exposed; physical and chemical state of the substance; the toxicity of the substance to man or experimental animals that can be related to man; and the amount of the substance likely to be found in other sources such as food and air. For example, because organic mercury is known to be relatively high in certain marketable fish taken from the sea, but not high in most drinking water, the tentative limit for mercury in water was adjusted downward to keep the total daily intake for man well below a hazardous level.

Perhaps a brief discussion of the mercury case may give some insight as to how criteria are developed and standards selected. No limit was set for mercury in the 1962 Drinking Water Standards, because it was not considered a very likely contaminant of drinking water. After all, since the inorganic mercurials are not very soluble in water and not very well absorbed, it is not considered to be as toxic in water as other forms of mercury.

Nevertheless, during the 1960s, there was a decided change in attitude about the hazard of mercury in the environment. For example, the experiences at Minamata Bay, Japan, were brought to light and more recently, the occurrence of the relatively large concentration of organic forms of mercury in the food chain in this country have caused great concern. The USPHS set a tentative limit of 0.005 mg/l for mercury in drinking water in 1970, even though there was little specific data for this particular exposure. At that time, there was considerable difficulty with analytical methods for mercury; much resampling and splitting of samples between laboratories was necessary to establish how much mercury there really was in various media and in which chemical form. Because of the general distribution of mercury, however, the Federal Technical Committee for reviewing standards went ahead and developed a standard for drinking water.

Review of current information indicates that among populations exposed to fish containing methyl mercury, the minimum effect level occurs when the blood methyl mercury level is 0.2 mg of mercury per gram of blood. This value corresponds to prolonged, continuous exposure to approximately 0.3 mg/day of methyl mercury for a man. Applying a safety factor of 10 to this minimum effect exposure level, the maximum allowable exposure by all routes should be 30 mg/day. This allows for a maximum daily consumption of 60 g of fish at the current FDA guideline of 0.6 mg/kg of mercury with no other exposure.

Although there is health criterion for total intake, there is also a need to allocate a certain sensible portion to all routes of exposure–food, water and air. In 1970, the technical committee decided that 0.005 mg/l could be allotted to water. With a 2-liter consumption, this would be a third of the maximal allowable intake of 30 mg/day.

In the meantime, more has been discovered about the occurrence of mercury in drinking water and about the removal rate by conventional water treatment. Of 273 water supplies sampled only one exceeded 0.002 mg/l; thus, apparently, mercury in drinking water is not a major problem. The single sample showing an excess of 0.002 mg/l was extensively resampled and the elevated concentration did not occur again. In addition, several of the conventional water treatment processes will reduce the mercury content of the water. Inasmuch as the mercury concentrations are relatively high in certain foods and low in drinking water (now measurable below 1 mg/l) the Public Advisory Committee thought the mercury standard for water should be 0.002 mg/l. The limit would be based on total mercury and might be considered as a screening level. More elaborate analyses would be performed on samples in excess of the standard to determine the proportions of alkyl and inorganic form.

Incidentally, the extent of toxicological data will help determine if a factor of safety should be 2, 10 or as high as 500. In the case of certain inorganics with a long history of human health observations, 2 or 10 may be satisfactory, but in the case of chlorinated hydrocarbon pesticides where the use has been recent and only experimental animal data are available, a factor of 500 may be necessary. In certain cases, toxicological data and epidemiological observations do not necessarily support the same conclusion; so wherever there is a reasonable doubt, the conservative limit is used until more conclusive evidence is accumulated. Because the arsenics presently fall into this category, a discussion of the circumstances may serve to illustrate how the committee copes with such a situation.

In 1962, the mandatory limit of 0.05 mg/l was not changed, but a recommended or preferred limit was set at 0.01 mg/l, because it was thought that arsenic may be carcinogenic. During the recent review in 1970, the technical committee used several consultants, and one actually believes that the public suffers from arsenophobia. He contended that toxicity studies have established the usefulness of arsenic as a poultry feed additive without adverse effects and that the overly restrictive limits could be relaxed.

Considerable animal toxicological data exist on the effects of inorganic arsenite and arsenate, and minimum-effect and no-effect levels in dogs, rats and mice have been determined. Three generations of breeding mice were exposed to 5 mg/l of arsenite in the diet with no observable effects on reproduction. The fact that little of the organic form of arsenic found in shrimp meat is retained was demonstrated in rats.

Nevertheless, in considering a health effect limit for arsenic, there are conflicting data to consider. Admittedly, attempts to demonstrate through animal studies that arsenic is tumorigenic have met with failure; chronic toxicity studies indicate that mg/l concentrations can be consumed without effects; and furthermore, occupational health studies have demonstrated that man can be exposed to rather high concentrations without adverse effects.

On the other hand, clinical and epidemiological studies conducted in several locations in the world have demonstrated adverse health effects of arsenic in drinking water. For example, in an area on the southwest coast of Taiwan where the population uses water from wells with high arsenic content, some people develop hyperpigmentation, keratosis and, finally, skin cancer. These conditions increase with age, which of course indicates longer exposure, but the effect is more pronounced in males. The prevalence rates for skin cancer show an ascending gradient according to the arsenic content of the well water. The conditions of melanosis, keratosis or skin cancer were not noted in control villages where the arsenic content was 0.02 mg/l or less. In the study villages, these conditions were noted, but were much less prevalent when the arsenic content was less than 0.3 mg/l.

Another example is from Antofagasta, Chile, where the water supply from 1958-1970 contained 0.8 mg/l arsenic and dermatological manifestations were noted, especially in children. The water supply of this city of 130,000 inhabitants has been changed by treatment, and Dr. Jose Manuel Borgono from Chile has stated that these adverse conditions are not noted when the water supply contains less than 0.15 mg/l of arsenic.

A third field example comes from a ranch in Nevada where well water with an arsenic content that ranged from 0.5-2.75 mg/l is reported to have resulted in a skin condition in one child and possibly in another.

Total intake and retention are usually difficult to estimate in such studies because arsenic is a geochemical pollutant and can be expected to occur in the air, food and water of those sections of the country where it occurs naturally in soil and water in higher concentrations. Therefore, much more research is needed on distribution and availability of arsenic in various media before any major change is made in the standards. Nevertheless, because of the interpretation given to field observations by epidemiologists, the present recommendation is to limit arsenicals in water to 0.05 mg/l.

One other factor that influences limits is attainability. This certainly is demonstrated in the area of coliform bacteria, which can be easily controlled with a disinfectant without really knowing precisely what the factor of safety is for a certain occurrence or level of coliform.

Many utilities and surveillance programs have relied on chlorine to eliminate pathogens and minimize the sanitary indicator, coliforms, in the distribution system. Experience has taught that in certain situations, the chlorine residual is a reasonable indication of the absence of pathogens, and thus is a partial substitute for the usual sampling for coliform bacteria. A proposal has therefore been made that this test be so used where the chlorine demand is minimal, where residuals have traditionally been maintained, where the turbidity is consistently kept below 1 Formazin Turbidity Unit (FTU), and where the coliforms are below 1/100 ml and general bacterial population is less than 500/ml.

General organic content or odorous limits may also be determined by the available technology and by just how much the community wants to spend for a high-quality water. For example, before the 1962 Standards were published, small activated carbon filters were used by USPHS researchers in Cincinnati to capture and characterize some organics in water that were retained on the surface of this type of carbon. They were thus able to demonstrate that in certain surface sources there were many potentially dangerous or odorous chemicals. Based on a large national sampling, a limit of 0.2 mg/l was selected as the limit for organics, detected by extracting the carbon with chloroform (CCE) as a means of curbing excessive odors or organic concentrations of an ill-defined nature. A vast majority of supplies were below that limit.

Inasmuch as the limit in the 1962 Standards is not a mandatory, but merely a recommended one, not many waterworks supplies have been monitored routinely for this parameter. Obviously, because a more practical method of gauging for harmful organics is necessary, we have been doing water examination for both specific and groups of chemicals. The results to date indicate we may suggest a limit for trihalomethanes (such as chloroform), which are formed when certain waters are chlorinated.

Sometimes a criterion for a constituent limit is such that the analytical work can be reduced by including the effect of the substance with some other generalized limit. This is the case with phenol, where a limit of 1 mg/l was selected in 1962, because more than that amount tended to react adversely with chlorine to form components. The 1970 EPA Technical Review Committee recognized that the analytical methods for the concentration have not been very practical for most waterworks people and, therefore, decided to drop phenol from the standards. The odor limit serves as the means to control the phenol inasmuch as it is that indirect effect that is significant.

The historical development of limits for pesticides may also serve to illustrate how care and caution have to be used in setting limits for materials added to the environment for presumably beneficial reasons. In 1962, considerable effort was made to examine the logic and need for including selected limits for specific or general groups of pesticides in the Drinking Water Standards. Finally, a decision was made to wait and accumulate more information on simplified analyses and mammalian experiments before setting limits. In 1965, another committee was formed to advise the USPHS on effectively implementing the Drinking Water Standards' suggested limits for pesticides, but they were never formally adopted for enforcement. Although these limits were based mainly on fish toxicology studies, they did serve as guidelines in evaluating a water supply and in controlling excessive use of herbicides in reservoirs. Since then, many investigators have developed experimental animal data and made some observations of human experiences, so a different set of limits has been proposed in the current revision. Nevertheless, the factors of safety in some cases still have to be as high as 500 because of incomplete information regarding long-range effects of chlorinated hydrocarbons.

Generally speaking, any standard is made up from a set of criteria, plus an action plan to implement the desired goal; therefore, developing and carrying out a monitoring or surveillance program is an integral part of the standard. In the case of pesticides or radionuclides, regional characterization of the sources, types and levels by state or federal programs may help minimize the analytical work for the local water utility. Variances in this part of the standards will probably be permitted on a case-by-case basis. Certainly, historical constituent level patterns, manufacturing operations, or land-use practices will help state and federal engineers decide what a continuing monitoring schedule should be for compounds that man has added to the environment.

Water analysts will want to also discuss other sampling schedules with the state administrators and field engineers, because laboratory space, equipment and personnel will depend on their requests. Common sense must be used in establishing and then modifying sampling frequencies with time and experience.

For example, as mentioned previously a certain experience may allow the chlorine residual test to be substituted for some of the usual coliform samples in the distribution system. The same principle of variance may apply to many chemical constituents. Generally, where the population is greater and the system larger, more samples should be taken to establish the quality of water at the consumer's tap.

There are some unresolved matters regarding organics standards, but we hope the constituents included and the limits for each have been discussed and understood so that we can proceed with a surveillance program. In the meantime, we in the EPA continue to try to develop the health effects data and control technology necessary to improve the scientific basis for and attainment of the standards.

14

SETTING STANDARDS FOR WASTEWATER EFFLUENTS–PRESENT STATUS AND FUTURE TRENDS

Kenneth M. Mackenthun
Director, Criteria and Standards Division
Office of Water Planning and Standards
U. S. Environmental Protection Agency
Washington, D. C.

PAST TRENDS

Before the present and future are addressed, often it is well to assess the past. The programs and activities of the present and future are influenced by the successes or failures of those programs and activities previously implemented. Water pollution control programs tend not to be advanced in a leap-frog fashion with periods of acceleration interspersed with periods of maintaing status quo. Rather, they are advanced through a steady and persistent evolutionary effort to formulate better programs that can be feasibly implemented to attain society's clean water goals.

Past efforts in water pollution control have been associated with the development and universal implementation of methods to cleanse domestic sewage prior to its introduction into receiving waterways, in an effort to reduce the threat of communicable diseases, oxygen-demanding materials and suspended solids associated with such wastewaters. Following these efforts, society's attention was directed toward reducing the harmful qualities of industrial wastes. Our present and future concerns appropriately are being directed to toxic, carcinogenic and other health-related wastes, and diffuse wastes and their effects from nonpoint source areas. The thrust of water pollution control has evolved from a focus directed principally toward a single entity of the problem to a focus on the control of all aspects of the total water pollution problem.

Prior to 1972, federal involvement in water pollution control was limited principally to enforcement when the health or welfare of persons in one state was affected by pollution from another state; to the approval of state water quality ambient standards; to technical support of state water pollution control activities; and to research, development and demonstration of water pollution control methods and procedures. In 1972, it became clear to Congress that such activities were inappropriate to deal with the accelerating water pollution problems and that society would have to exert a greater and more comprehensive effort to stem the tide of water pollution.

PRESENT TRENDS

On October 18, 1972, the Federal Water Pollution Control Act Amendments of 1972 became Public Law 92-500. Further, they became a landmark in the history of environmental legislation and embodied an extremely complicated, comprehensive and highly ambitious program to clean the nation's waters. The thrust of this legislative package was focused on new federal concepts of effluent limitations, regulation of toxic substances, and a means to ensure that effluent limitations would attain, maintain and protect water quality, as well as to continue and accelerate the implementation of the water quality standards program initiated by federal law in 1965.

Sections 301, 304, 306 and 307 of the 1972 Act mandate the development and promulgation of effluent limitations for existing industrial sources, standards of performance for new sources, and pretreatment standards for new and existing sources. Point sources of discharges of pollutants are required to comply with these regulations, where applicable, under the provisions of section 402 of the Act–the National Pollutant Discharge Elimination System. The term point source was defined in the Act as meaning any discernible, confined and discrete conveyance including, but not limited to, any pipe, ditch, channel, tunnel, conduit, well, discrete fissure, container, rolling stock, concentrated animal feeding operation, or vessel or other floating craft, from which pollutants are or may be discharged.

The Environmental Protection Agency was mandated to promulgate uniform regulations that may be applied to individual permits and enforced in court without the necessity of treating each discharger as a unique situation. In this endeavor, consideration must be given to factors that may result in different limitations for different plants. For example, such factors would include the age and size of the plant, raw materials used, manufacturing processes, products produced, available treatment technology, energy requirements and costs. In its effluent guidelines development effort, the Agency is required to define the attainable levels for each industry by the application of the best practicable control technology currently available, as well as the application of the best available technology economically achievable. Thus, such guidelines are technology-based effluent limits that are developed following an appraisal of the treatment technologies available to a particular industrial category and the economic considerations associated with the installation of such treatment technology for the industrial group. In its past substantial efforts in the promulgation of effluent guidelines, the Agency has considered principally the conventional sanitary engineering wastewater constituents such as biochemical oxygen demand, chemical oxygen demand, total suspended solids, pH and similar constituents. In some cases, effluent limits have been placed on other appropriate wastewater constituents.

An industry can discharge either directly to a stream, in which case appropriate effluent limitations are applied, or it can discharge to a publicly owned treatment works in accordance with treatment standards defined pursuant to section 307 of the Act. Pretreatment standards are designed to prevent the discharge into publicly owned treatment works of any substances incompatible with the treatment operation. The current regulations relating to pretreatment have caused confusion in the implementation of this program, and the Agency currently has revisions under consideration that should resolve current problems in the pretreatment area.

Virtually all substances are toxic to humans or to other organisms if present in water in sufficient concentrations over extended time periods. Certain substances are so highly toxic and impose such severe risks to human or aquatic life that special regulatory mechanisms to control their presence in the aquatic environment may be warranted. For such substances, section 307(a) of the Federal Water Pollution Control Act Amendments of 1972 creates procedures designed to control such pollutants. Section 307(a) mandates the publication of a list that includes any toxic pollutant for which an effluent standard will be established. Within 180 days following the publication of the list, the Agency is mandated to publish proposed effluent standards for the pollutant, which shall take into account its toxicity, persistence, degradability, the usual or potential presence of the affected organisms in any waters, the importance of the affected organisms, and the nature and extent of the effect of the toxic pollutant on such organisms. Following the proposed publication, a formal public hearing must be held within 30 days, and not later than six months after the publication of the proposed effluent standard. The standard shall be promulgated unless the administrator finds that a modification of such proposed standard is justified based upon a preponderance of evidence adduced at the hearing. The Agency is promulgating final regulations under section 307(a) to control the direct discharges of DDT, aldrin/dieldrin, endrin, toxaphene, benzidine and polychlorinated biphenols. These standards will be implemented through the National Pollutant Discharge Elimination System program. Pretreatment standards will be proposed under section 307(b) of the Act for the above-named standards.

The concept of management of water pollution through the implementation of water quality standards was continued in section 303 of the 1972 Act, which provides that states shall adopt water quality standards that consist of the designated uses of navigable waters involved and of the water quality criteria for such waters based upon such uses. Such standards shall protect the public health and welfare, enhance the quality of the water, and serve the purposes of the Act.

Water quality standards are an essential part of the state's water quality management system:

1. They publicly define the state's water quality objectives and form the basis for its planning;
2. They serve as a basis for determining the National Pollutant Discharge Elimination System permits;
3. They provide a basis for effluent limitations for pollutants not specifically addressed in the effluent guidelines or for pollutants for which the effluent guidelines are not stringent enough to protect desired water uses;
4. They serve as a basis for evaluating and modifying management practices for the control of nonpoint source wastes;
5. They serve as a basis for judgment in other water quality-related programs such as water storage for the regulation of stream flow, water quality inventories, control of toxic substances, thermal discharges, cooling lakes, aquaculture, and dredge and fill activities; and
6. They contain the state's antidegradation policy.

In implementing this section of the Law, the applicable regulation (40 CFR 130.17) provides guidance to the states in formulating an antidegradation policy, in upgrading water quality standards to meet the 1983 goals of the Act, and particularly, in mandating that prior to any downgrading of existing water quality use designations, the state must demonstrate that such uses are unattainable because of natural background water quality, irretrievable man-induced water conditions, or because the application of effluent limitations more stringent than best available treatment economically achievable would result in substantial and widespread adverse economic and social impact. The national water quality goals expressed in the 1972 Act stated that the discharge of pollutants into the navigable waters be eliminated by 1985 and that, wherever attainable, water quality that provides for the protection and propagation of fish, shellfish and wildlife, and for recreation in and on the water, be achieved by July 1, 1983. State water quality standards now are being reviewed and revised to establish use classifications and water quality criteria necessary to protect public health and to achieve the national water quality goals for all waters of the United States.

Water quality criteria are an integral part of water quality standards. Such criteria define the levels of quality constituents necessary to maintain specified water uses. The genesis of water quality criteria in the United States occurred in the early 1900's with the publication of the effects of various types of waste on fish (Marsh, 1907; Shelford, 1917). M. M. Ellis (1937) of the U. S. Bureau of Fisheries produced a monumental early effort to describe and record the effects of various concentrations of a great number of substances on aquatic life. He reviewed the existing literature for 114 substances in a 72-page document and listed lethal concentrations found by various authors. In addition, he produced a rationale for use of a standard test animal in aquatic bioassay procedures and made excellent use of the common goldfish and the water flea in such experiments. Following the work of Ellis, efforts were made to standardize bioassay procedures in the scientific literature, particularly in the use of fish as test organisms.

In 1952, the state of California published a 512-page book entitled *Water Quality Criteria* that contained 1369 references. This 1952 edition was tremendously expanded and enhanced into a second edition edited by Jack E. McKee and Harold W. Wolf and published by the Resources Agency of California, State Water Quality Control Board (1963). This edition included 3827 cited references and combined under one cover the world's literature on water quality criteria as of the date of compilation. Such criteria were identified and categorized according to their effects on domestic water supplies, industrial water supplies, irrigation waters, fish and other aquatic life, shellfish culture, and swimming and other recreational water uses. The results of such a tabulation presented a range of values and, as would be expected by those investigating such conditions, there often was an overlap in values between those concentrations that had been reported by some investigators as being harmless to aquatic organisms and those concentrations that had been reported by others as being harmful under the conditions of the experiments to the same organisms. Such an anomaly is due to differences in investigative techniques, the characteristics of the water used as a diluent for the toxicant, the physiological state of the test organisms, and variations in temperature under which such tests were conducted.

In 1966, the Secretary of the Interior appointed a number of nationally recognized scientists to a National Technical Advisory Committee to develop water quality criteria for specified water uses. A book emanated, which constituted a comprehensive documentation on water quality requirements for particular and defined water uses (1968). In some respects, this volume represented a marriage between the best available experimental or investigative criteria recorded in the literature, and the judgments of recognized water quality experts with long experience in associated management practices. They produced recommended criteria for concentrations of water constituents to ensure the protection of the quality of the aquatic environment and a continuation of the designated water uses. This book was enlarged and revised by the National Academy of Sciences and the National Academy of Engineers in a publication entitled *Water Quality Criteria, 1972*, a 592-page report of the Academy's Committee on Water Quality Criteria (1974).

The latest in the saga of water quality criteria is an EPA publication dated July 26, 1976, entitled *Quality Criteria for Water*, which addresses about 60 water constituents and contains a concise recommended criterion, a brief introduction to the pollutional qualities of the constituent, a rationale that depicts principal concentration-effect relationships relating to the development of the recommended criterion, and a list of references cited. Such recommended criterion are designed to protect public health, water uses, the aquatic and other organisms that live within or depend upon water for their existence, and the consumers of such aquatic or other life.

Section 302 of the 1972 water Act provides that where the application of the best available technology economically achievable would still interfere with the attainment or maintenance of water quality that would (a) protect public water supplies, agricultural and industrial uses, (b) protect and propagate a balanced population of shellfish, fish and wildlife, and (c) allow recreational activities in and on the water, more stringent effluent limitations shall be established, which can reasonably be expected to contribute to the attainment or maintenance of the above-designated uses. This is a statutory means of effecting water pollution controls more rigorous than national standards for specified waters, but in so doing, the economic and social costs of achieving such controls, the economic or social dislocation in the affected community or communities, and the reasonable relationship between the economic and social costs and the benefits to be obtained must be recognized.

FUTURE TRENDS

The future often is inextricably intertwined with the present, and so it is with water pollution control. The future is embodied in a settlement agreement signed on June 7, 1976, by representatives of the Environmental Protection Agency, the Natural Resources Defense Council, Inc., and the Citizens for a Better Environment. This agreement provides that the EPA shall develop and promulgate regulations that will (a) establish and require achievement, at the earliest possible time but in no case later than June 30, 1983, of effluent limitations and guidelines for classes and categories of point sources, (b) require application of the best available technology economically achievable for such category or class, and (c) result in reasonable further progress toward the national goal of eliminating the discharge of all pollutants, including toxic pollutants.

Pursuant to the settlement agreement, the EPA is committed to a program to investigate 65 designated pollutants that represent substantial concern related to their human health or environmental effects (see Appendix A). Twenty-one major categories of industries were identified to receive priority consideration for tighter effluent limitations (see Appendix B).

The agreement provides for the development of human health and ecosystem-related environmental criteria that will define the need for regulatory control of the substance of concern and for the concentration within the aquatic environment that will protect human health and the aquatic resource. The development of such criteria will address individually each of the 65 designated substances of concern and is scheduled to result in an appropriate publication of such water quality criteria by June 30, 1978. The information associated with this developmental effort will serve as the basis and purpose for the Agency's regulatory effort to control the discharges of these substances into the aquatic environment. Many of the 65 substances of concern are recognized human carcinogens, mutagens, or teratogens, or are potential candidates for such labels. Likewise, many of those substances are organic substances which generally have not been addressed in effluent guidelines or in quality criteria for life in water, and thus will represent a new era both in the development of effluent guidelines and in the development of water quality criteria. The 65 designated substances were selected because of their known occurrence in point source discharges to the aquatic environment; their presence in fish or in drinking water; their evidence of carcinogenicity, mutagenicity, or teratogenicity; the likelihood of human exposure; and their persistence, accumulation within the biotic foodweb, and toxicity (acute and chronic) to aquatic organisms and others that may feed upon such aquatic life.

In addition to the development of appropriate criteria for each of the 65 designated substances of concern, the Agency will be determining their known and potential release into the environment, treatment technology applicable to the control of the release of such substances into receiving waterways, and the economic constraints associated with such control efforts. The explicit emphasis in this program will be on human health and environmental effects, and the program will require treatment technology studies and control efforts on an industry-by-industry basis and a consideration of health and the environmental effects of the 65 substances on an individual constituent basis.

This integrated strategy makes it unnecessary to enforce stringent, low-risk standards for a handful of pollutants while doing virtually nothing about many others. By using the industry-by-industry approach, the strategy takes into account the economic effects of the controls that are to be imposed. By providing an assessment of the total distribution of each toxic substance before attempts are made to regulate it, the strategy ensures that the decision-makers will have a clear idea of the pathways by which the pollutant enters the water environment and a better chance to manage them.

The results of implementing this integrated strategy will be to develop effluent limitations to ensure the use of the best available technology economically achievable to reduce the level of toxic substances entering the water environment. The regulations developed pursuant to the agreement are to establish effluent limitations and guidelines, new source standards of performance, and pretreatment standards for each of the 21 classes or categories of industries not later than December 31, 1979. The implementation of the effluent limitations developed will be principally through the National Pollutant Discharge Elimination System's permit. If it should be determined that the reductions provided by appropriate treatment technology are insufficient to protect water quality, the option is available to tighten the controls, pollutant by pollutant, through the section 307 mechanism of the 1972 water Act or through section 302, which provides (paragraph 12 of the settlement agreement) that not later than June 30, 1978, the administrator shall establish a specific and substantial program to: (a) determine whether more stringent effluent limitations, guidelines and standards are necessary to attain and maintain water quality in a specific portion of the navigable waters; (b) ensure protection of public water supplies, agricultural and industrial uses, and the protection and propagation of a balanced population of shellfish, fish and wildlife; and (c) allow recreational activities in and on the waters.

REFERENCES

Ellis, M. M. "Detection and Measurement of Stream Pollution," *Bull. U. S. Bur. Fish.* 48:365-437 (1937).

Marsh, M. C. "The Effect of Some Industrial Wastes on Fishes," Water Supply and Irrigation Paper No. 192, U. S. Geological Survey, pp. 337-348 (1907).
McKee, J. E. and H. W. Wolf. *Water Quality Criteria,* Pub. 3-A (Sacramento, California: State Water Quality Control Board, 1963).
National Academy of Sciences and National Academy of Engineering. *Water Quality Criteria* (Washington, D. C.: U. S. Government Printing Office, 1974).
National Technical Advisory Committee. *Water Quality Criteria*, a report to the Secretary of the Interior (Washington, D. C.: U. S. Government Printing Office, 1968).
Shelford, V. E. "An Experimental Study of the Effects of Gas Wastes Upon Fishes, With Especial Reference to Stream Pollution," *Bull. Ill. State Lab. Natural History* 11:381-412 (1917).
Water Quality Criteria (Sacramento, California: State Water Pollution Control Board, 1952).

APPENDIX A
SPECIFIED POLLUTANTS

Acenaphthene
Acrolein
Acrylonitrile
Aldrin/Dieldrin
Antimony and compounds
Arsenic and compounds
Asbestos
Benzene
Benzidine
Beryllium and compounds
Cadmium and compounds
Carbon tetrachloride
Chlordane (technical mixture and metabolites)
Chlorinated benzenes (other than dichlorobenzenes)
Chlorinated ethanes (including 1, 2-dichloroethane, 1, 1, 1-trichloroethane and hexachloroethane)
Chloroalkyl ethers (chloromethyl, chloroethyl and mixed ethers)
Chlorinated naphthalene
Chlorinated phenols (other than those listed elsewhere; includes trichlorophenols and chlorinated cresols)
Chloroform
2-Chlorophenol
Chromium and compounds
Copper and compounds
Cyanides
DDT and metabolites
Dichlorobenzenes (1, 2-, 1, 2-, and 1, 4-dichlorobenzenes)
Dichlorobenzidine

Dichloroethylenes (1, 1- and 1, 2-dichloroethylene)
Dichloropropane and dichloropropene
2, 4-dimethylphenol
Dinitrotoluene
Diphenylhydrazine
Endosulfan and metabolites
Endrin and metabolites
Ethylbenzene
Fluoranthene
Haloethers (other than those listed elsewhere; includes chlorophenylphenyl ethers, bromophenylophenyl ether, bis (dischloroisopropyl) ether, bis- (chloroethoxy) methane and polychlorinated diphenyl ethers)
Halomethanes (other than those listed elsewhere: includes methylene chloride, methylchloride, methylbromide, bromoform, dichlorobromomethane, trichlorofluoromethane, dichlorodifluoromethane)
Heptachlor and metabolites
Hexachlorobutadiene
Hexachlorocyclohexane (all isomers)
Hexachlorocyclopentadiene
Isophorone
Lead and compounds
Mercury and compounds
Naphthalene
Nickel and compounds
Nitrobenzene
Nitrophenols (including 2/4-dinitrophenol, dinitrocresol)
Nitrosamines
Pentachlorophenol
Phenol
Phthalate esters
Polychlorinated biphenyls
Polynuclear aromatic hydrocarbons (including benzanthracenes, benzopyrenes, benzofluoranthene, chrysenes, dibenzanthracenes and indenopyrenes)
Selenium and compounds
Silver and compounds
2, 3, 7, 8-Tetrachlorodibenzo-*p*-dioxim (TCDD)
Tetrachloroethylene
Thallium and compounds
Toluene
Toxaphene
Trichloroethylene
Vinyl chloride
Zinc and compounds

As used throughout this Appendix A, the term compounds shall include organic and inorganic compounds.

APPENDIX B
INDUSTRIAL POINT SOURCE CATEGORIES

1. Timber Products Processing
2. Steam Electric Power Plants
3. Leather Tanning and Finishing
4. Iron and Steel Manufacturing
5. Petroleum Refining
6. Inorganic Chemicals Manufacturing
7. Textile Mills
8. Organic Chemicals Manufacturing
9. Nonferrous Metals Manufacturing
10. Paving and Roofing Materials (Tars and Asphalt)
11. Paint and Ink Formulation and Printing
12. Soap and Detergent Manufacturing
13. Auto and Other Laundries
14. Plastic and Synthetic Materials Manufacturing
15. Pulp and Paperboard Mills; and Converted Paper Products
16. Rubber Processing
17. Miscellaneous Chemicals
18. Machinery and Mechanical Products Manufacturing
19. Electroplating
20. Ore Mining and Dressing
21. Coal Mining

15

REGULATORY AGENCY EVALUATION OF PRESENT POSITION OF THE WATER AND WASTEWATER INDUSTRY RELATIVE TO STANDARDS– Viewpoint of the State Regulatory Agency for Water Supply

William A. Kelley, P. E.

Chief, Division of Water Supply
Bureau of Environmental and Occupational Health
Michigan Department of Public Health
Lansing, Michigan

Reason is an absolutely essential ingredient in the federal water supply program, standards and regulations if the Safe Drinking Water Act is to be implemented in a meaningful way. Unfortunately, I am convinced that this will be accomplished only if an alert waterworks industry– state, local and private–insists upon it.

To understand the EPA thinking behind the standards, it is necessary to briefly review two small but important sections in the Act. The word contaminant is defined as "any physical, chemical, biological, or radiological substance or matter in water." In other words, anything other than pure H_2O is a contaminant when present in water. In addition, Section 1401 (1) (B) indicates that the National Primary Drinking Water Regulations should "specify contaminants which *in the judgment* of the Administrator, *may* have *any* adverse effect on the health of persons" (emphasis added). In considering this particular section, it appears that the door has been left wide open for EPA to enact controls and standards on any type of contaminant that may be present in water whether or not it is a known health concern.

Most persons associated with the public drinking water supply programs around the country are familiar with the fact that the original draft of the National Interim Primary Drinking Water Regulations, published early in 1975, was totally unacceptable and in no way considered costs. These proposed regulations and standards were not reasonable and would have been totally unworkable for the states and water utilities. The March 1975 draft was extensively revised to provide the interim regulations as they were promulgated in December of 1975. The promulgated Interim Regulations are not perfect, but certainly are substantially improved over earlier drafts. The waterworks industry did influence EPA in finalizing these interim regulations.

We now must anticipate EPA's revision of the interim standards along with other portions of the water supply program and their publishing of the Revised Primary Drinking Water Regulations during the next year. The subject of chloroform is reviewed in considerable detail in this book. A representative of the National Cancer Institute states that facts as they are known today do *not* demonstrate that chloroform in the concentrations present in drinking water is carcinogenic to humans, but representatives of EPA say there *will* likely be a standard for chloroform or, more likely, for the entire family of trihalomethanes.

Asbestos has also been discussed at this seminar in some detail. Statements have been made acknowledging that asbestos fibers inhaled from air are recognized carcinogens. However, to date there are no facts to support the contention of some persons and groups that asbestos-like fibers are carcinogenic when ingested in drinking water. Still, a representative of EPA has indicated that there should be some type of standard for asbestos in drinking water. Further, it was stated that since EPA had forced the citizens of Duluth to install complete treatment on their water supply, in like manner it also should be forcing San Francisco and Seattle, among others, to make necessary changes in their treatment facilities. This is not a very responsible method of establishing standards for protection of the health of persons.

A representative of the National Cancer Institute indicated that although food and air pollution are probably more serious sources of chloroform than drinking water, it is easier to study chloroform in water. The other sources are too abstract and complicated at this time to study effectively in detail. Therefore, the National Cancer Institute is studying the concentration of chloroform in water supplies. The results are talk about the fact that drinking water contains chloroform, and speculation as to carcinogenicity. The subject must be studied to determine whether chloroform in drinking water is a hazard. However, a problem develops for the water supply industry when some persons and self-appointed consumer groups take the position that *all* or almost all the chloroform must be removed from drinking water regardless of cost. It appears that EPA, or some EPA staff members, are somewhat in sympathy with these organizations, at least on the subject of chloroform in drinking water.

This type of thinking demonstrates that EPA does not plan to maintain a reasonable drinking water program, a cost-effective program truly aimed at protecting the health of humans. It appears that there are some theorists who are controlling the thinking and policy of EPA and are attempting to force the expenditure of millions of dollars in requiring strict controls where it is not known that *any* controls are necessary. These are the areas in which the waterworks industry must be watchful. As a group, we must consider all such proposed requirements, and we must make our thoughts known to EPA and our representatives in Congress so that mandated programs be established and carried out on a reasonable basis. Otherwise, standards may be developed in the courts or be unduly influenced by special interest groups rather than be formulated by knowledgeable public health professionals.

In Michigan, the state program is making progress. Act 399 of the Public Acts of 1976 has been adopted by the Legislature and on January 4, 1977, was signed into law. This is Michigan's Safe Drinking Water Act and basically provides statutory authority to bring the Michigan program in line with the requirements of the federal act. As a part of Act 399, Michigan has adopted by reference the water quality standards of the National Interim Primary Drinking Water Regulations, which were in effect when Act 399 became law. These standards are in effect and will remain so until state drinking water standards are adopted following promulgation of the National Revised Primary Drinking Water Regulations.

I trust that the water supply industry will remain cognizant of EPA activities and carefully monitor the Revised Drinking Water Regulations and national program. That is essential if we are to maintain an ongoing water supply program at state and local levels. New regulations and quality standards must be developed with reason, consistency and in a way that can be supported by the waterworks industry in general. We at the state health department need public assistance in reviewing any EPA standards and regulations concerning drinking water, and encourage people to make their comments known not only to us but to EPA and their representatives in Congress.

16

REGULATORY AGENCY EVALUATION OF PRESENT POSITION OF THE WATER AND WASTEWATER INDUSTRY RELATIVE TO STANDARDS– Viewpoint of the State Regulatory Agency for Wastewater

Paul L. Blakeslee

Regional Engineer
Water Quality Division
Michigan Department of Natural Resources
Lansing, Michigan

In discussing the topic we need to begin with an understanding of the context in which the term standards is used. A strict definition implies establishment, by an authority, of a rule for measure of quality. To be a standard, there must be authority for establishing the rule.

At the federal level, this authority originates in federal statutes for both water and wastewater. At the state level, authority derives from Act 245 of 1929 as amended, and is vested in the Michigan Water Resources Commission.

State water quality standards have existed since the late 1960s and are currently in the process of revision. Proposed revisions were developed by departmental staff and have been made available for public hearing by action of the Water Resources Commission. Hearings were held in November and December and further refinements, based on the hearing record, will be considered for adoption by the Water Resources Commission.

Staff of the Department of Natural Resources and Michigan Department of Public Health are currently drafting the first State Water Quality Standards under Act 245 authority for ground waters of the state. Consideration of this segment of the state's water resource, using the format of water quality standards, will bring a new dimension to many aspects of the state's environmental program.

Water quality standards are the only true standards within the state's water management program. There are, however, many other criteria and limitations that the Water Pollution Control Industry encounters on a day-to-day basis. Some of these are prescribed federal criteria in the form of industrial category performance standards or treatment definitions such as the Secondary Treatment criteria for municipal facilities. Criteria for a specific discharge will be developed by departmental staff and may be set forth in the form of discharge permit limitations. In general, the limitations will reflect the more restrictive of the federal standards, or performance necessary to meet the state's water quality standards.

Standard-setting agencies walk a tight rope between those interested in use of water as either a disposal medium or renewable resource, and those who view it as potential water users or who seek to apply strict environmental protection standards.

Some groups look with a relatively narrow perspective at the quality criteria to be established for a given water. The perspective taken by Michigan must be one of equal evaluation of all identified interest and impacts.

As the technology of investigation, analysis, data evaluation and assessment of consequence of action are better defined, we are constantly looking at new forms of alphabet soup (CN, Hg, PCB, PBB, Merex, Penta). As we face these challenges, we must evaluate the relative risks of various levels of such constituents in the many elements of the environment.

Water quality standards are at the central point of two distinct ecosystems: (a) the natural system of plants, birds, animals (including man) associated with use of waters; and (b) the economic ecosystem that combines resources, energy and production for the benefit of man in an ever-shrinking world.

Visualize two equally industrious spiders, one representing environment and one representing economics, building webs with strands connected to each side of the water quality standards issue. There are many individual threads in those interconnecting webs that, when tightened or loosened, cause a shift of dozens of other elements on both the economic and the environmental sides of the web. The role of water quality standards should be to balance the economic and environmental interests so that the most symmetrical (balanced) webs possible can coexist.

After reviewing the hundreds of pages of testimony and documents submitted in response to public hearings held on proposed revisions to the State Water Quality Standards, one staff member recently commented that a certain proposed revision "must be satisfactory—the number of responses indicating that the proposal was too restrictive about equaled the number of responses indicating that it was not restictive enough."

It is not an ideal way to operate, but possibly that is where the standard-setting process is today. Some of this opposite pole reaction, or tugging at the threads by the two spiders, is due to a lack of full understanding of what the effects are on the other side of the web.

When high-quality treatment is demanded to produce one benefit, is there recognition that additional range land in Montana may be mined for coal to produce the power required to operate the treatment facility? When any methyl ethyl compound is proposed to be discharged to Lake Ontario, is there recognition that recreational and commercial fishing may be jeopardized?

There is continuous improvement in our technological ability to ask the right questions. It is up to us—the economic sector, the environmental sector and the regulatory agencies—to come up with the right answer.

Reluctance to accept a prescribed standard or criterion is often based on a feeling that adequate information was not considered in the establishment of the criterion. This may be an accurate assessment in some cases. In many current situations, adequate facts are not available, and the best current judgments must be used. In such cases, the criteria are established according to the facts available at that time and must be either confirmed with additional facts or modified as time and experience show a better way. Criteria set with less than the desired amount of information may be either too liberal or too conservative, and only if there is an ongoing questioning and searching for additional answers will the correct balance be established.

We will probably never reach the point where yesterday's answers are correct for tomorrow's questions. The best we can do is answer today's questions on the basis of today's information, looking at the issue in the broadest possible perspective, and implement the program that comes from those answers.

The more issues that can be freely discussed and resolved on the basis of discussion, the fewer confrontations we will face as we work together toward the common objective of water quality protection.

17

REGULATORY AGENCY EVALUATION OF PRESENT POSITION OF THE WATER AND WASTEWATER INDUSTRY RELATIVE TO STANDARDS– Toxicology and Trace Contaminants Standards

Joseph K. Prince

Regional Toxiciologist and Health Effects Specialist
U. S. Environmental Protection Agency
Region V
Chicago, Illinois

Only within the last 10 years have instrumentation capabilities and microchemical techniques become sophisticated enough to allow analysis and measurement of trace quantities of contaminants in the environment. In drinking water alone, more than 1200 organic contaminants have been reported. Further analysis led to the realization that many of these chemical agents are suspected or known carcinogens. It has been recognized that the presence of such trace organics in concentration of 1 ppm or lower could adversely affect aquatic organisms. The number of compounds that have been identified is related to those new limits of detection and the analyst's capabilities; these are now in the realm of 10^{-2}-10^{-1} mg/l for drinking water.

Pollution is not something new, but recently the number of hazardous chemicals has grown quite large. Because of these large numbers of potential carcinogenic and mutagenic agents, the toxicologist has come to the foreground and is now an important member of the scientific advisory team that helps to ensure that the effects of environmental contamination on the public health receives primary consideration.

Protection of public health is still the most important task facing science. To help carry out this function, the Safe Drinking Water Act were legislated by Congress. The Environmental Protection Agency was charged with the responsibility of setting standards and controlling types and amounts of chemicals that could be discharged into the environment. With the advent of this new responsibility came the realization by the public and private sectors of our society, of the role toxicology must play in serving these functions. It is known how toxicology has been applied to acute situations but now comes the question of the effects of long-term, low-level exposures. Recent investigations of some 6000 chemical compounds have produced more than 1000 chemical agents that have carcinogenic activity. There are countless others that have mutagenic or teratogenic capabilities or are plainly toxic or lethal. Knowledge of the various types and concentrations of organic material in the environment would be most valuable in helping to assess the problem of health-effects screening. Selecting specific compounds is most difficult in this maze of chemicals, since very little chronic or long-term information is available on dose-response relationships. Add the difficulty of extrapolating lab test results of lower animals to human health effects, with variable and/or unknown concentrations of trace contaminants and you begin to understand the difficulty in providing exact answers to the various questions posed by interested individuals and impacted by recent legislation.

Which chemicals should be regulated? What information should you gather? What criteria should be used to regulate chemicals in the environment? Is a carcinogenic reaction a singular response of the species tested, or is the species particularly susceptible or immune? Thousands of questions confront toxicological evaluation, and concern mounts to higher levels.

Formerly, we were most concerned about acute effects. Presently, the main question focuses on the chronic long-term, or delayed health effects of human exposure to these organic constituents in our environment. An assessment of human susceptibility can be made, based on the careful evaluation of present bioassay studies and/or techniques. I would like to stress the words assessment (as opposed to absolute conclusions) and careful evaluation (as opposed to trying to directly correlate lower animals and humans). As a toxicologist I am interested in the susceptibility of biological cells, organisms and/or systems to exposure from chemicals that are exogenous or normally outside that system, as opposed to internal metabolic pathways or biochemical function.

Modern toxicology, as we know it today, was born just over 100 years ago in the mind of M. J. Orfila, a Spaniard from the Island of Minorca. He studied chemistry and mathematics, and later medicine, and is known as the Father of Modern Toxicology because of his interests, which centered on harmful effects of chemicals as well as therapeutic effects.

He also introduced quantitative methodology into study of actions of drugs and chemicals on animals. Two most important word concepts are toxic (harmful) effects versus therapeutic (beneficial) effects. Most people put them on opposite sides of the fence. However, they are not opposites. They are both part of the same dose-response curve, which tells us what biological response will occur when a given dose of a chemical is applied to an organism, cell, organ system or enzyme pathway that has a receptor site. It is this effect or response that is most important to humans in preventing adverse effects on our health.

Adverse effects from chemicals have been known since the first dose of poison was given. As a toxicologist, I have verified hundreds of human deaths from toxic doses of various chemicals. Overdoses of alcohol, barbiturates, narcotics of all sorts, tranquilizers, and their potent combinations, are common knowledge to the toxicologist. Research goes on with thousands of chemicals in many testing laboratories, using animals to define effective toxic or lethal doses of chemicals; the physiological responses elicited; the adverse or beneficial effects; and how the physiology of a biological organism is affected by this chemical.

Research is carried on to define the essential nature of chemicals in their relationships to biological systems. The toxicologist aims to define the adverse effects that are manifested when such chemicals enter into a biological system. In acute situations, when large doses have entered a biological system and toxicity and/or fatality occur, physiological examination can usually indicate such a condition. However, what happens when, due to a sublethal dose, no such condition manifests itself over a protracted period of time? Or is there a response, which is not observed? It is also possible that there is a chronic effect, which will manifest itself after a latent period. At each level of dosage we must ask ourselves: Will this dose induce adverse effects? What are the adverse effects I am willing to risk? There are more questions than answers.

As a toxicologist, I do not have an adequately developed body of knowledge on which to make absolute determinations of risk/benefit analyses. Standards are promulgated based on reason, logic, established scientific principles, regard to potentially sensitive populations, rates of reactions, chemical kinetics, knowledge of biochemical pathways, pathology, physiology and a host of other materials. They are generally defined in a rational manner, based on the consideration of large variables mentioned, and with public health as the primary objective. But how does one set standards for trace contaminants in our drinking water, which only recently were proven to be present due to greater resolution of our detection ability. How do we set human exposure standards when information on chronic sublethal exposures does not exist? Certainly we cannot adjust standards based on new-found abilities of detection in the analytical labs. Basic research is, of course, the primary need. To relate directly to the human is essentially impossible because of variation between and within species. However, if one considers that all these test organisms and the human belong to the mammalian system, and are composed of similar cell and organelle functions, then some relationships, however subtle, might be detected and correlated. Since we have no human data to work with, it is a starting point. (Human data are occassionally seen when accidental deaths or suicides occur as a result of chemicals. This pathological information is then included in data evaluation.) Test data gathered for medicinals is evaluated on such a basis, using biological systems, which are relatable in terms of importance to the animal Man.

Dogs are used to evaluate lung-dose responses, cats for the central nervous system, and monkeys for organ and muscle systems. Qualitative data is ascertained for inferential judgments. This type of data has given us many beneficial medicinals, valuable to today's society.

Penicillin to combat infection, and antibiotics, were boons to mankind, as is diabenase and orinase for diabetics, nitroglycerine for heart patients, and procaine for the sensitive dental patient. Each is a treasure for man's benefit, good health and comfort, but only when employed in limited dosage! If you take too much, pathological changes can occur that are the toxic manifestations of man's body chemistry and can cause injury or death. Barbiturates, phenothiazines and valium, etc. are most important psychopharmacological tools when used properly, but overdoses are frequent. All these chemicals were tested on animals and the results extrapolated for human use. Imagine if we had to wait an extra 25 or 30 years for penicillin or the corticosteroids and many other of the beneficial medicinal chemicals. I am not advocating throwing caution to the wind. On the contrary, we can use those same principles we have used for analyzing those medicinal chemicals (caution and care in our consideration, using tried and true concepts). Extrapolation of dose-response curves outside the range of experimental results and with animal data can be hazardous and is subject to inaccuracies because of lack of uniform response of a substance's toxicity for different mammalian species. But this factor was taken into consideration, and a risk/benefit assessment was made. Very careful evaluations took place.

Risk assessments of each of the chemicals available to human systems, were made on quantity received (dose), duration of exposure, human experiences, animal data and sensitive population. Of all the factors involved, dose (human) is by far the most important, because with each level of dose comes the relative effect. To paraphrase Paracelsus, dose alone is the characteristic that makes any chemical a poison. Only when a cell, organ or biological system can show its clinical effects can we relate quantity and quality of responses. Based on these considerations, judgments can be made and standards set. We know it may be imprecise, but it is reasonable and necessary.

Lack of available data on relative toxicity of chemicals of statistically significant or sensitive populations, data from various animal species, dosage from the environment, exposure duration, biotransformations, genetic information, population mobility and background noise are just some of the obstacles facing toxicologists in the standard-setting function. But we must provide the best judgment available for any given situation.

We have come, therefore, to another realization of life. Toxicologists are not absolute and definitive scientists such as engineers, mathematicians and physicists. We are, at once, practitioners of an art, integrating information received at many levels and from various directions, remaining sensitive to the needs of a society and concurrently following the allowed principles of scientific endeavors. We are to fulfill tasks set before us by governmental and political bodies and, concurrently, are sometimes viewed with disdain by the very element we try to serve. To provide absolute numbers is in the realm of the aforementioned professions. The toxicologist's role is the safety and health of the public—to provide the best route available toward maintaining a high level of public and environmental health and, concurrently, to carry on liaison with research seeking those answers that will provide us with more definitive data on which to make judgments for setting standards. In an effort to provide the best advice available, judgments will be made that may have to be modified at a later date due to new information. Flexibility is a most important factor to consider, for that characteristic allows for change—a vital necessity. If there is no room for flexibility and for accompanying change, if standards are set too rigidly, or attempted to be set by structured rules, the rules will be broken. As knowledge increases, flexibility will allow us to modify our rules toward more precise standards. But until then, Science Advice is a real and necessary programmatic function in standards-setting.

The National Academy of Sciences has the heavy burden of recommending health standards for the Safe Drinking Water Act, and they are the only body that can carry out this function. From the outset, the Health Effects Committee knew that the legislative mandate of an "absolute guarantee of health safety from any known or anticipated adverse effects from chemical contaminants" could not be done for all or even most of the contaminants studied. Therefore, with respect to human health, the Committee said:

1. Where sufficient data of human dose-response relationship can be projected with some degree of precision, a projection will be made. This projection will be explained and its qualifications made explicit.
2. For contaminants where data are of sufficient quantity and quality, the NAS will exercise its scientific judgment and identify and propose contaminant levels for which it anticipates the risk of adverse health effects to be specifiable and very small. The risks at the proposed levels will be described with an explanation why no "safe" level has been identified.
3. For contaminants for which evidence provides no scientific basis or methodology for recommending levels, the Academy will describe available data and its significance in terms of known or anticipated adverse health effects.

To me this attitude of tread carefully but progressively is much more reasonable and safer than providing ad hoc committees, laws and regulations after the damage has been done. Advice is available and reasonable, but must be utilized to be of significance or benefit.

Science advisor mechanisms are now beginning to play a significant role in national and state legislation and programmatic input. The state of Michigan and its governor should be commended that they are leaders in implementing Science Advice at the state level.

Science Advice is reasonable and necessary. It is progressive and moderated by caution and good sense. It acts within it own parameters and thus is valid. It can be modified or amended, and is not so restrictive as to break if stress is applied. It can survive in short- or long-term situations.

It is a tool with which to make decisions and judgments before significant adversity is produced. It is the basis of our function in EPA in carrying out the regulations of the Safe Drinking Water Act and Toxic Substances Acts. We will carry out this function to the best of our ability, striving to prevent significant adverse health effects to our public and environment. But we need the understanding of all, and to realize we are human and have limitations.

SECTION V

OPERATIONS EXPERIENCE

18

PRETREATMENT AND TRACE CONTAMINANTS

Michael Garnell

Supervisor of Filtration
Detroit Metro Water Department
Detroit, Michigan

In recent years the American people's attitude toward their public water supplies has been increasingly critical. The general, widespread acceptance of and confidence in the quality of these supplies is diminishing rapidly and apprehension and distrust are becoming more prevalent. Water purification plant operators are made painfully aware of this trend by the increased incidence of complaints and the tendency to blame the water for all kinds of disorders.

The industry had previously earned the confidence of the consumer by virtually eliminating all waterborne diseases caused by pathogenic organisms. The few complaints received were not for any concern for the safety of the drinking water but rather objections regarding its aesthetic qualities. Now we are confronted with a different type of pollutant. These modern contaminants do not produce any dramatic disease after a short incubation period. Instead, they exert a prolonged environmental stress, the cumulative effects of which are suspected of causing some of the afflictions of middle and old age and may even accelerate the aging process itself.

Regaining the level of consumer confidence previously enjoyed will be extremely difficult if not impossible. Most people are unwilling to tolerate even traces of a suspected toxicant in their drinking water. The apprehension caused by the specter of all kinds of allergens, mutagens, carcinogens and teratogens in the drinking water may be creating more psychological than physiological stress. These fears can no longer be allayed by continual reassurances that the water is free of pathogenic bacteria. Today, people are worried about cancer, not typhoid. They are asking,

"What have you done for me lately?" As long as harmful pollutants are suspected to be present in drinking water, consumers are tempted to purchase devices or gadgets that claim to purify water.

Unfortunately, the conventional water treatment plants in operation today are not equipped or designed to cope with modern pollutants. Activated carbon, the only treatment chemical extensively used that is capable of removing soluble contaminants, has been employed primarily to combat tastes and odors. The massive dosages that would be required to remove pollutants in the parts per billion range are neither practical nor cost-effective.

Disinfection practices have evolved with the objective of providing an adequate factor of safety against biological diseases. After the germicidal efficiency of free chlorine was discovered to be greatly superior to chloramines, breakpoint chlorination became the standard practice. When water became implicated as a possible vehicle for the transmission of the more resistant viruses, treatment plant operators were persuaded to carry higher and higher free chlorine residuals. These practices are now being questioned because of the production of chlorinated organics.

The question is, can our conventional pretreatment processes be changed in our existing plants in a way that will ameliorate the present problem of trace contaminants? Recent experiments have demonstrated that the concentration of these undesirable chloro-organics can be greatly reduced by delaying the application of chlorine until after the turbidity is removed, strongly suggesting that the particulate matter is part of the source of organics or bears some relation to it. In addition to the suppression in the plant of undesirable biological growths that prechlorination affords, most operators will be reluctant to forego the added control of the disinfection process and final finished water residual that it provides.

A possible substitute for the loss of control that results when postchlorination alone is employed would be the diligent use of accurately performed laboratory chlorine demand tests. In 1951 Feben and Taras made some detailed studies of the chlorine demand of Detroit water and many nitrogenous compounds. This notable work elucidated the overall behavior of chlorine with chlorine-consuming compounds such as ammonia, amino acids and proteins. Their findings demonstrated a definite relationship between the chlorine demands at different contact periods that can be expressed mathematically. This mathematical relationship makes it possible to to regulate chlorine dosages to provide a desired specific chlorine residual in the finished water for any future time. By plotting chlorine demand against contact time on log paper, a straight line should result. The general form of the equation of this curve is:

$$D_t = kt^n$$

where D = chlorine demand at time t
k = one-hour demand
t = contact time in hours
n = slope of the line or tangent

By using the half-hour and one-hour demands, the equation becomes

$$D_t = D_1 \, (D_1/D_{1/2})^{3.32 \log t}$$

where D_1 and $D_{1/2}$ are the one-hour and half-hour chlorine demands, respectively. The value of the exponent n in the basic equation provides a clue regarding the speed of the reaction. The lower the value of n, the more rapidly the chlorine is consumed. Inorganic oxidizable ions, such as ammonia and sulfite, would show n values approaching zero, whereas complex organics such as protein, which are slowly oxidized, would have the highest values.

If postchlorination only is to be employed, the treatment plant operator must know what dosage to apply. Chlorine demand tests on the raw water will give values that include the effect of the turbidity, which can be substantial. Therefore, the raw water to be tested should be coagulated and filtered to a degree of clarity comparable to the plant effluent before the chlorine demand test is performed. Determinations of the half-hour and one-hour demands can then be used to calculate dosage requirements for the plant filtered water necessary to produce a safe residual in the distribution system.

It must be emphasized that chlorine demand tests require meticulous care in the preparation of solutions and glassware and in the measurement of chlorine residuals, in order to be of any value as a procedure for controlling chlorine dosage. Many plants may not have personnel sufficiently skilled in analytical procedures and techniques to perform the tests with adequate accuracy and reliability. Also, a water whose chlorine demand fluctuates unpredictably over a wide range will require frequent testing. Raw water from Lake St. Clair, which is probably more stable than many other surface supplies, requires chlorine dosages ranging from less than one to more than three parts per million. In any case, the elimination of prechlorination will cause some loss of control of residuals. It would be imprudent to introduce any changes in treatment that risk a resurgence of the old diseases just to reduce the chlorinated organics only *suspected* of being a problem. Although most people are not worried about typhoid, operators must continue to be concerned about bacterial contamination.

If the turbidity *is* the source of much of the chlorinated organics, then every effort should be made to minimize the turbidity of the finished water. There is probably no phase of the pretreatment process that offers a greater potential for improving the clarification process at a low cost than rapid mixing. The initial mixing process has been grossly neglected in the design and operation of many of our water purification plants. The result has been an unnecessary waste of coagulant chemicals and sometimes less than satisfactory turbidity levels in the filter effluent. Operators accustomed to running jar tests are aware of the significant role the initial mix plays in the overall clarification process. Maximum mixing efficiency results when complete homogenization of the coagulant and raw water stream is achieved within a few seconds. If this objective is not realized quickly, much coagulant will be wasted and part of the colloidal particles will not be destabilized, resulting in higher filtered water turbidity. The evidence seems to indicate that the critical reactions between the coagulant and the colloidal particles take place rather quickly and are irreversible.

The Water Works Park Plant in Detroit has no mechanical mixing equipment, and the only mixing provided is a baffled chamber. A considerable improvement was achieved by using a pressurized alum solution diffuser with numerous orifices covering the entire cross section of the raw water stream. Although this provides the very desirable multipoint application that expedites dispersal, it does not provide the more essential ingredient of turbulence that the colloid destabilizing process requires. What is needed is a multipoint application at or followed immediately by an adequate velocity gradient without any short-circuiting or backmixing. The expense of the added head loss, which a good rapid mix would necessarily incur, would be more than compensated for by the increased efficiency of the coagulation process.

Most large surface water treatment plants have been designed with only two main objectives in mind—clarification and disinfection. It may be possible to reduce trace contaminant levels to some extent by modifying and improving these processes. However, the degree of improvement in quality wanted by many consumers and the complete removal of contamination clamored for by some environmental groups cannot be realized. It would be reasonable to attempt to effect this degree of purification for that portion of the water we ingest, but not for the billions of gallons that pass through our treatment plants today.

REFERENCE

Feben, D. and M. J. Taras. "Studies on Chlorine Demand Constants," *J. Am. Water Works Assoc.* 43:922 (1951).

19

SOME EXPERIENCES WITH LOW-LEVEL OZONATION

Wilfred L. LePage
Water Superintendent
Monroe, Michigan

This chapter is intended to be a nontechnical account of some experiences encountered during an investigation into the application of ozone to potable water treatment at Monroe, Michigan. These experiences were both good and bad; there was cause both for jubilation and for despair. Nonetheless, confidence in ozone as a useful tool for potable water treatment has solidified over the years and a new, 450-lb/day ozone plant is nearing completion.

The primary objective in that plant will be the destruction or reduction of objectionable odors in a municipal drinking water supply treating the waters of the western basin of Lake Erie—waters not noted for pristine purity.

Ozonation was investigated because it offered a safe, easily controlled system of proven reliability that promised to do the job without objectionable aftereffects. That, however, was before the energy crunch when the term trihalomethane existed only within the vocabulary of the organic chemist.

In 1973, investigative work progressed in an atmosphere of optimism and enthusiasm. Water plants in Canada were visited and their successes with ozone were noted with interest. An ozonator was leased and a 50,000-gpd pilot plant was placed on-stream. The effects of 1-5 mg/l of ozone applied to raw water and to partially treated water (chlorinated, treated with alum and settled) were carefully observed (LePage, 1974).

The first disappointment occurred when it became obvious that the pilot plant data with regard to odor destruction might not be totally conclusive. This was ascertained because as the pilot plant was readied to go on-stream, the odor intensity in the raw water diminished and still has not returned to a seriously high level. This is attributed to unusually high lake elevations, but as the lake level declines, the more intense odors will probably recur. However, the data obtained indicated that up to 1.5 mg/l ozone would effectively destroy all routinely occurring odors, some of which reach threshold values of up to 8 or 10. The data from this study, combined with reports of the successful use of ozone for the same purpose at other locations with similar or more severe problems, suggested that odor control through ozonation could also be attained at Monroe.

Most other parameters monitored yielded predicted and favorable results (LePage, 1975). Briefly, the most gratifying results appeared in areas quite important to potable water treatment and are as follows:

1. effective destruction of all odors appearing during operation of the pilot plant with no objectionable secondary odor formation;
2. rapid and effective disinfection;
3. improved coagulation reactions (with alum) with regard to time, floc size and settling characteristics; and
4. the apparent ability of ozone to destabilize certain difficult-to-treat, colloidal waters, making them more amenable to alum treatment.

Of secondary interest, but no less gratifying, were:

1. ready destruction of the small amounts of color sometimes appearing in the raw water;
2. some turbidity reduction following ozonation with no apparent increase of suspended solids;
3. rapid oxidation of nitrite to nitrate;
4. ammonia destruction under certain conditions (above pH 9.5);
5. some COD reduction; and
6. effective and rapid destruction of cyanide.

Contrary to prediction and still somewhat a mystery, was the behavior of chlorine demand following ozonation. It was predicted that the ozone would essentially remove all chlorine demand from the water. This proved to be untrue. Experiments repeatedly indicated that chlorine demand following low-level ozonation (1-5 mg/l) decreased only slightly. As more chlorine was added to measure the demand, a scattering effect, or ripple, interrupted the curve, and chlorine demand actually increased. In many instances the chlorine demand following ozonation was greater than it was before ozonation. Further chlorine demand reduction could only be attained through the application of unrealistically larger doses of ozone. Unfortunately, this effect occurs within the desired and proposed range of ozonation at Monroe.

Little was made of this at the time. It was discussed by several people in the ozone field, and generally dismissed as interesting but not really important. That, however, was in 1973, and the whole possibility of a potential health threat existing from chloroform and other trihalomethanes produced in our drinking water following chlorination had not yet surfaced. But some years later, EPA researchers investigating the use of ozone for the removal of trihalomethane precursor (Love *et al.,* 1976) encountered what appears to be essentially the same effect. They report as follows:

> "Although ozone alone would not form trihalomethanes, low-level ozonation followed by chlorination produced as much (or more) chloroform as with chlorination alone. This means that the potential for trihalomethane formation was not reduced by low-level ozonation and that subsequent chlorination to produce a disinfectant residual in the distribution system would result in trihalomethane production."

Thus, low-level ozonation appears largely ineffective for removal of chlorine demand and, as reported by the EPA, ineffective for the removal of trihalomethane precursor. In fact, it sometimes increases chlorine demand and enhances trihalomethane formation. Why? No one is sure. Love *et al.* (1976) theorize:

> "Ozone may alter some material that would not normally participate in the haloform reaction as a precursor. It could also be possible that because the ozone satisfies some of the oxidant demand, more chlorine is available for the haloform reaction."

This appears to be supported by the chlorine demand tests at Monroe. The ripple in the chlorine demand reduction curve occurs within the range of ozonation required for disinfection and odor reduction which is also the point at which an ozone residual develops. The development of an ozone residual implies that the immediate oxidant demand is satisfied, and subsequent addition of chlorine may initially be consumed in the haloform reaction (Figure 1). Or, since it is known that even low-level ozonation exerts a chewing-up effect on some of the heavier molecules, they probably become more reactive to chlorine creating higher chlorine demand and producing more trihalomethanes.

Whatever the reason, this was probably the greatest disappointment encountered during the entire ozone project. Unfortunately, these developments occurred after the Monroe pilot plant was dismantled, and follow-up work was not possible. However, after start-up of the new ozone plant in mid-1977, full-scale investigations will be conducted to determine the most effective application of the many treatment options available in the very versatile Monroe plant, in order to minimize trihalomethane formation and produce the very highest quality water possible.

$$\begin{matrix} & & CH_3 \\ \diagdown & & \diagup \\ & C{=}C & \\ \diagup & & \diagdown \end{matrix} + O_3 \rightarrow O{=}C \begin{matrix} CH_3 \\ \diagup \\ \\ \diagdown \end{matrix}$$

$$\begin{matrix} CH_3 & \\ \diagdown & \\ & C{=}O \\ \diagup & \end{matrix} + Cl_2 \rightarrow \begin{matrix} CCl_3 & \\ \diagdown & \\ & C{=}O \\ \diagup & \end{matrix}$$

$$\begin{matrix} CCl_3 & \\ \diagdown & \\ & C{=}O \\ \diagup & \end{matrix} + Cl_2 \text{ or } (OCl^-) \rightarrow HCCl_3 + -\overset{\overset{O}{/\!/}}{C}-OH$$

Figure 1. A probable reason for increased chlorine demand following ozonation.

Today's high construction costs are rarely a cause for rejoicing. Rather, they are a fact of life that must be accepted with the satisfaction that those high costs can be met. Ozone plants are no exception.

The cost of constructing ozone treatment facilities will vary with each installation. The quality of the water to be treated and the treatment objectives will dictate the ozone equipment capacity. The type of ozonators best suited for the specific application, the degree of automation and amount of instrumentation desired, and the design of the contacting system can vary the initial capital investment and significantly influence the cost of operation. The cost of the complete ozone treatment facility at Monroe and all the yard piping required to connect to the existing treatment works is as follows:

Earthwork	$ 5,800
Bearing caissons	26,000
Reinforced concrete tank and substructure	111,000
Superstructure housing	39,000
Ozone equipment and instrumentation	270,000
Ozone distribution and recycling	25,000
Piping and valves	235,000
Total	$711,800

These figures represent the price to the city, including the contractor's markup, but excluding engineering and design.

The only raw materials required for the production of ozone are electricity, a little cooling water (which is recovered) and air. Therefore, close attention has been given the recent escalation of electric power costs and the amounts required for operation of the ozone generators. Most manufacturers of ozone-generating equipment quote power requirements of about 12 kWh/lb of ozone. Our job specifications included that figure. At that time, in 1973, Monroe Waterworks was buying electricity at just under 2¢/kWh. This meant that at 1973 prices, ozone would cost about 23¢/lb. Today, the net cost of power is about 2.5¢/kWh and this would boost the cost of ozone to about 30¢/lb were it not for the outstanding efficiency of the Monroe ozonators (manufactured by PCI Ozone Corporation, West Caldwell, New Jersey). Under actual test conditions conducted by Monroe personnel, the Monroe machines produced ozone at a continuous output 17% above their rated capacity at a total power consumption of 8.6 kWh/lb of ozone or 28% less than job specifications required. This included operation of all auxiliary equipment under maximum power-consuming conditions. This means that at today's escalated power costs, ozone production will cost 1.5¢/lb less than anticipated three years earlier. This, indeed, was cause for rejoicing!

The cost of operating an ozone plant should be only slightly more than the cost of the electric power it consumes. Today's ozone equipment is reliable, long-lasting, may be automatically controlled and is reasonably undemanding for maintenance. There is no labor involved for handling ozone and no space required for storage, and concurrent savings of other treatment chemicals through the supplementary benefits of ozonation help offset the initial capital investment.

These are just some of the more outstanding experiences in the Monroe ozone project. In summary, the Monroe Water Department remains enthusiastic about ozonation with respect to control of objectionable odors and to to the many important and useful supplementary benefits that occur simultaneously.

REFERENCES

LePage, W. L. "Ozone Studies at Monroe," presented at the Thirty-Fifth Annual Meeting, Michigan Section, AWWA, Boyne Falls, Michigan, (September 17-20, 1974).

LePage, W. L. "Ozone Treatment for Monroe, Michigan," *Proceedings of The Second International Symposium on Ozone Technology,* Montreal, Quebec, Canada, The International Ozone Institute, (1975).

Love, O. T., Jr., J. K. Carswell, R. J. Miltner and J. M. Symons, "Treatment for the Prevention or Removal of Trihalomethanes in Drinking Water," in *Interim Treatment Guide for the Control of Chloroform and Other Trihalomethanes,* Appendix 3 (Cincinnati, Ohio: United States Environmental Protection Agency, 1976).

INDEX

INDEX